T0172987

Thin-walled Cellular Structures

(METHODS FOR THEIR ANALYSIS)

V.A. Ignatiev and O.L. Sokolov
Volgograd Civil Engineering
and Architecture University
Volgograd
Russia

Foreword by
William E. Saul
Michigan State University
East Lansing, MI
USA

A.A. BALKEMA/ROTTERDAM/BROOKFIELD/1999

Authorisation to photocopy items for internal or personal use, or the internal or personal use of specific clients, is granted by A.A. Balkema, Rotterdam, provided that the base fee of US$1.50 per copy, plus US$0.10 per page is paid directly to Copyright Clearance Center, 222 Rosewood Drive, Danvers, MA 01923, USA. For those organizations that have been granted a photocopy license by CCC, a separate system of payment has been arranged. The fee code for users of the Transactional Reporting Service is: 90 5410 281 0/99 US$1.50 + US$0.10.

ISBN 90 5410 281 0

A.A. Balkema, P.O. Box 1675, 3000 BR Rotterdam, Netherlands
Fax: +31.10.2400730; E-mail: balkema@balkema.nl
Internet site: http://www.balkema.nl

Distributed in USA and Canada by
A.A. Balkema Publishers, Old Post Road, Brookfield, VT 05036-9704, USA
Fax: 802.276.3837; E-mail: info@ashgate.com

Foreword

This monograph in the field of **plate or thin walled structures,** translated from Russian, offers new and valuable material, much of which has not been available previously in English. Although specifically written to address the problems of complex or very large plate or thin-walled structures, a brief review of the contents reveals procedures and numerical methods of much more general interest and applicability, such as extracting a specified number of accurate eigenvalues for a complex numerically modeled structure. Persons working in steel, other metals, reinforced concrete, fiber reinforced polymers, or wood panels, for example, with civil structures including bridges, buildings and tanks; aeronautics and astronautics including air- and spacecraft; transportation including automotive and rail structures; marine architecture including coastal structures, ships and undersea vessels; defense; academics and researchers with an interest in plate structures, and many others—perhaps those working with mico- or nano- electronic or mechanical structures or in the bio or medical field— will find this book valuable. The possibilities seem endless in this era of rapidly breaking technology where developments specific to one application appear valid for another entirely different use.

Numerical methods, with special emphasis on the finite element method (FEM), have, in conjunction with the fantastically fast arithmetic (number crunching) capabilities of the digital computer, rendered a revolution in solid mechanics. The design and analysis of structures and machines has been unalterably changed forever. Our capacity for better, different, more wonderful and advanced designs are now only constrained by our capacity to conceive advanced designs, our ability to model what we conceive, and the tools with which we are able to address our design to the computer. Therein we are indeed challenged by the gulf between the theoretical or academic side of the FEM and its implementation, its practical usage. Great strides are being made in input preparation and output presentation that narrow that gulf daily; however, the modeling of complex and very large structures as well as input of its parameters and the algorithms which form the heart of the internal computations to render a solution to provide design parameters remain very much a challenge.

The strength of this monograph lies in its presentation of modeling techniques and numerical methods for the solution of very large or complex structures. Each item begins with basic, well-known theory and methodically develops procedures useful in application. Several alternative procedures are usually presented along with example problems designed to illustrate the procedure while highlighting details necessary for the user.

The major themes contained in this monograph related to two major approaches: One is the utilization of the finite element method (FEM), by substructuring for static analysis and design or for the computation of a specified subset of characteristic values (the always formidable problem of computing eigenvalues and eignevectors, at times called the eigen-problem) necessary for the solution of problems of stability or vibration. The second approach is the generation of functions which describe the overall behavior of certain types of plate structures without recourse to the FEM and its attendant necessity for superposing on the structure a network of artificial "elements" (in place of members) with attendant nodes, each with a number of kinematic degrees of freedom, usually six, which anchor the resultant large set of linear equations. With n-kinematic degrees of freedom there will be n-linear equations with n-unknown values; the displacement method of analysis will yield n-unknown displacements, the flexibility method will yield the same number of generalized forces. However, the latter method is applicable and limited to, as stated, only certain types of structures, those which meet the criteria of Chapter 3, such as a cellular bridge deck.

The FEM, with its very general applicability and great power, is generally the method of choice for structural analysis for all but the simplest applications. Thus substructuring, along with matrix condensation and superelements, the topic of Chapter 1, is a powerful tool for reducing an overwhelming problem into tractable size parts. At six degrees of freedom per node a finite element model with 10,000 nodes would produce 60,000 simultaneous linear equations; thus the solutions provided herein are designed for models with 10^3 and more, say 10^5 degrees of freedom. Chapter 1 is devoted to substructuring and introduces some innovative methods such as spline interpolation. In simple terms, the structure is modeled in a stepwise fashion from the elementary finite element to a cluster of elements, hopefully a recurring cluster, into a group of clusters—an elementary level superelement, again hopefully recurring, and so on with more and more complex superelements until the entire structure is the resultant superelement. The process following for static, stability, or dynamic analysis sweep forward to form the basic element proceeding through as many steps as deemed necessary from the first superelement to the most complex, i.e., the structure as a whole, for which case the analysis is performed. The process then performs a reverse sweep or return during which desired values are computed from the earlier obtained results, e.g., displacements, until values are obtained as desired, perhaps forces and displacements for each basic element or frequencies for the first m-modes.

Chapter 2 provides several innovative methodologies for the analysis and design of statically loaded thin-walled or plate structures. The approach is very general to allow for almost any configuration and the presentation is extensive so as to provide alternatives and worked-out example problems. There is extensive introduction to the use of spline interpolation in the methodology presented as well as algorithms for implementation. The material herein should be most valuable for designers for static loads.

The characteristic value problem, the solution of which provides eigenvalues and eigenvectors, nearly always for a limited subset, is discussed in depth in Chapter 4. Several innovative solutions, provided for the first time in English, are presented along with worked solutions. This is an extensive study of major concern in the dynamic or stability analysis of large or complex structures. Herein, the discussion is in terms of thin-walled structures; however, the numerical methods are equally applicable to most problems in the vibrations of solids or the stability of structures. The eigenvalues yield the natural frequencies of the structure and their associated vectors yield the corresponding mode shapes. The parallel stability problem yields the load multipliers and the buckling shapes respectively. For practical purposes it is usually sufficient to obtain less than 10% of the possible characteristic values, frequently much less. For example, with 10^3 to 10^5 degrees of freedom no more than fifty, perhaps less than ten eigenvalues and vectors suffice. However, it is most desirable to obtain accurate solutions. This turns out to be a formidable challenge that is ably met herein.

I recommend this monograph as a valuable reference addition to your library.

Michigan State University
July 13, 1997

William E. Saul

Preface

Structures consisting of thin-walled cells and prismatic shells are extensively used in many fields today, especially in civil engineering and shipbuilding.

The advances made in the field of numerical techniques notwithstanding, analysis of such complex structures involves considerable effort. The prevailing methods can be tentatively divided into two groups: approximate methods based on conversion into a structurally anisotropic theoretical model of employing V.Z. Vlasov's theory of thin-walled members and numerical methods in the form of finite elements enhanced with the substructuring method.

The limitations of the approximate methods are quite evident.

Application of the finite element method to analysing such structures gives rise to a plethora of difficulties caused by the need to depict the structure in an analytical scheme, i.e., modelling, as a large number of finite elements, and thus the need to process the large volume of data generated, while considering the limitations of speed and the amount of memory required of even the latest generations of computers.

Development of the substructuring method has considerably lessened these difficulties. Nevertheless, the conventional form of substructuring analysis suffers from vital drawbacks, such as operation in several stages, necessitating storage of the stiffness matrices of substructures at all levels, and the limitations of the static condensation procedure.

The first drawback gives rise to immense computational effort consuming significant computer resources.

The second drawback is associated with the prohibition of blockwise elimination of the boundary nodes at the substructure nodes in the Gauss method due to discontinuities along displacements.

Aspects of analysing thin-walled cellular systems and prismatic shells have been examined in this monograph with special emphasis on the problem of developing efficient algorithms in terms of computer resources.

Chapter 1 deals with the substructuring method, the main types of condensation schemes for unknowns in the boundary nodes of substructures and partial cases of constructing the stiffness matrices of substructures.

Chapter 2 is devoted to the static analysis of thin-walled systems by the substructuring method. The author has proposed algorithms based on the concept of spline interpolation of the displacement fields in the forward and backward sweeps of substructuring analysis. Several examples discussed in this context point to the high efficiency and the immense possibilities of these algorithms.

Chapter 3 (co-author O.L. Sokolov) deals with analysing multiconnected plates and prismatic shells with a periodic structure widely used in various fields of technology (shipbuilding, construction of bridges and large tanks etc.). Analysis of such structures often represents an unusually complex mathematical problem whose approximate solution is based on changing to a structurally anisotropic theoretical model with moderate elastic parameters of the filler in the form of ribs. The field of application of such an approach is limited to cases of of a dense network of relatively weak ribs. This chapter offers a new approach to this problem based on the simultaneous use of V.Z. Vlasov's theory for analysing thin-walled members, based on the variational method and an extremely efficient mathematical system of equations for analysing periodic structures using the method of finite differences. This approach enabled achieving the maximum possible displacements of the mathematical aspects of the problem of analysing multiconnected plates and prismatic shells of periodic structures and, consequently, assisted in the organisation and solution of several new problems in the field of structural mechanics of shells: analysis of cellular structures made up of constituent members, analysis of constituent cylinders, analysis of pontoons and ships with a multiconnected section and oscillations of prismatic shells with a multiconnected section. Solutions to some of these problems are discussed in this chapter.

Chapter 4 holds pride of place in the monograph. It deals with condensation methods in problems of natural oscillations and stability. The differences in these problems vanish when their formulation is sufficiently abstract. The two problems converge into one, namely, computing the eigennumbers of quadratic matrices with real or complex members.

From the mathematical point of view, Chapter 4 has a single objective, viz., solving an incomplete problem of eigenvalues and eigenvectors of large, real symmetrical matrices. Such matrices have only real eigenvalues.

As fairly efficient methods and algorithms and programmes for implementing them have already been developed for small matrices, the attention of researchers is now focussed on large matrices. Some excellent methods have been developed to date but there is as yet no commonly accepted opinion regarding the advantage of one method over another.

Apart from intellectual satisfaction, there is also fiscal stimulus for accelerating progress as it influences the cost of computer time and programming.

The method of consecutive frequency-dynamic condensation propounded in Chapter 4 is a totally new method developed by the author and his students. The possibilities of its application run far beyond the range of structures studied in this book. In principle, this is a new method for solving incomplete symmetrical algebraic problems involving eigenvalues and eigenvectors. Various modifications of this method are examined herein for equations expressed in the form of displacements as well as equations represented in the form of the method of forces, i.e., the stiffness and flexibility methods respectively.

The examples analysed illustrate the computing efficiency of the proposed method and its modifications and the high degree of accuracy of the results.

It may be said that Chapter 4 helps to fill the present void in Russian and other literature of the last 20 years with reference to the incomplete algebraic problems of eigenvectors and eigenvalues.

The other chapters likewise contain much new material that to date has not been adequately dealt with in the literature.

This book should prove useful to many readers, but in particular scientific workers and research scholars engaged in theoretical aspects and methods of analysing spatial thin-walled structures. It ought also to be of use to structural engineers and applied mathematicians.

Contents

1

Method of Substructuring

1.1 Introduction

The application of the finite element method (FEM) to analysing complex spatial structures, as pointed out in the Preface, involves many difficulties because of the need to depict structures in the analytical scheme by a large number of finite elements and the consequent handling of a large volume of data.

Various suggestions to modify the FEM have been proposed to overcome these difficulties by reducing the volume of data input and storage in the computer memory, reducing the order of the system of equations and enhancing the computing possibilities of the program used for analysing complex structures.

Among the methods proposed, substructuring is the most efficacious, wherein the complex structure is represented as a set of substructures, each representing an aggregate of basic finite elements. In this approach, each substructure is defined in a convenient system of selected co-ordinates and analysed independently with boundaries common to the other substructures. This analysis provides a more compact and tractable stiffness matrix of the substructure and its nodal loading matrix. The substructure for which such matrices have been defined is generally called a superelement.

The system of equations written for the boundary nodes of the superelements expresses the equilibrium conditions of the entire structure as an aggregate of superelements. This system of equations contains far fewer unknowns than the basic FEM equations would have produced. Computation of displacement of the substructure nodes completes the first major set of calculations, which may be termed the forward sweep of calculations.

In the backward sweep of calculations, each substructure is analyzed for the given loading and the boundary displacements found in the initial computation. These calculations offer no particular difficulty since the substructures are invariably described by low order systems of equations. From the viewpoint of the classic method of displacements, each substructure (superelement) in such an approach represents a complex element of the main system of the displacement method (Sapozhnikov, 1980a, b; Przemieniecki, 1965, 1968).

When analysing very large or complex structures, representing them as aggregates of substructures of the same level may prove inadequate because of the high orders of the system of equations describing the analytical scheme of each substructure. The analytical scheme is therefore then organised in stages or levels. After that the complete structure is represented as an aggregate of hierarchic substructuring of several levels whose number is determined by the designed accuracy of resolution as well as the capacity of the computer system used.

The basic concepts of the substructuring method were first published by Przemieniecki (1963, 1965, 1968) and later extensively developed (Dashevskii and Rotter, 1984; Meisner, 1968; Postnov et al., 1975, 1979; Sapozhnikov, 1980a, b).

1.2 Constructing a Stiffness Matrix and Substructuring the Nodal Loading Matrix

Assume that the structure is divided into several substructures that can be regarded as superelements (SEs) of the n-th level. Any of these SEs may be divided into some number of SEs of the $(n-1)$-th level, $(n-2)$-th level etc. up to SEs of the first level consisting only of basic finite elements (zero level superelements).

When joining some superelements of the $(t-1)$-th level into a superelement of the t-th level, the equilibrium equations of the corresponding substructure can be represented as:

$$[K]\{q\} = \{p\}, \qquad \ldots (1.1)$$

where $[K]$ is the general stiffness matrix of the substructure comprising some superelements of the $(t-1)$-th level; $\{q\}$ is the vector of nodal displacements of the substructures; and $\{p\}$ is the vector of nodal forces acting on the substructure.

After classifying the nodal displacements (unknowns) into internal (slave or secondary) 'i' and external (boundary or main) 'e', eqn. (1.1) may be represented in the following block form:

$$\left[\tilde{K}\right]\{\tilde{q}\} = \{\tilde{p}\}, \qquad \ldots (1.2)$$

where $\left[\tilde{K}\right] = \begin{bmatrix} K_{ii} & K_{ie} \\ K_{ei} & K_{ee} \end{bmatrix}$ is the general stiffness matrix of the substructure

with renumbered unknowns; $\left[K_{ts}\right]$ are the stiffness matrix blocks corresponding to the reactions in node t to unit displacements of nodes s;

$\{\tilde{q}\} = \begin{Bmatrix} q_i \\ q_e \end{Bmatrix}$, $\{\tilde{p}\} = \begin{Bmatrix} P_i \\ P_e \end{Bmatrix}$ are the vectors of nodal displacements and nodal

forces with renumbered components; $\{q_i\}$ is the vector of internal nodal displacements; $\{q_e\}$ is the vector of boundary nodal displacements; $\{P_i\}$ is the vector of forces applied to the internal nodes; and $\{P_e\}$ is the vector of forces applied to the boundary nodes.

Here the internal (secondary) nodes refer to nodes in general whose unknowns are eliminated. Correspondingly, boundary nodes are all those nodes whose unknowns are retained in the analysis (boundary or basic nodes). In eqn. (1.2) the indices defining the number of the substructuring level have been dropped to simplify the expression.

By elimination from eqn. (1.2) of the vector of internal (slave) displacements of the substructure, we derive

$$\left[K_{ee} - K_{ei}K_{ii}^{-1}K_{ie}\right]\{q_e\} = \{P_e - K_{ei}\,K_{ii}^{-1}\,P_i\}, \qquad \text{... (1.3)}$$

or

$$\left[K_{ee}^{*}\right]\{q_e\} = \{P_e\}, \qquad \text{... (1.4)}$$

where

$$\left[K_{ee}^{*}\right] = \left\{K_{ee} - K_{ei}\,K_{ii}^{-1}\,K_{ie}\right\} \qquad \text{... (1.5)}$$

is the boundary stiffness matrix of substructuring of the t-th level; and

$$\left\{P_{e}^{*}\right\} = \left\{P_e - K_{ei}\,K_{ii}^{-1}\,P_i\right\} \qquad \text{... (1.6)}$$

is the loading vector of substructuring for the t-th level.

The process of elimination of the unknowns of group 't' in the system of equation (1.1) is called the static condensation of unknowns toward the boundary nodes of substructuring. Reduction in the order of the system of equations achieved in this manner is called 'condensation' or 'reduction' (Gallagher, 1984). The result is the conversion of the stiffness matrix and force vectors of the $(t-1)$-th level into the stiffness matrix force vector of the t-th level. This process commences with the zero level and is repeated until a substructuring stiffness matrix of the n-th level is attained, i.e., of the entire structure.

In the backward sweep commencing with a high level of substructuring, the displacements of all internal nodes of the superelement are found from eqn. (1.2) using the known boundary displacements:

$$\{q_i\} = \left[K_{ii}^{-1}\right]\{P_i\} - \left[K_{ii}^{-1}\,K_{ie}\right]\{q_e\}. \qquad \text{... (1.7)}$$

Using the general FEM procedure, all components of the stress state are calculated from the total displacement vector (1.7) for all the superelements of the zero level (basic finite elements).

1.3 Variants of the General Scheme of Condensation of Unknowns in Substructuring Boundary Nodes

1.3.1 *To Express Eqn. (1.1) in the Form of (1.2), the Rows and Columns in Eqn. (1.1) have to be Rearranged for Grouping the Unknowns into Internal and Boundary Nodes Separately.*

These rearrangements increase the band width of the stiffness matrix and as a consequence may significantly increase computation effort. To avoid this, an optimum renumbering of the internal nodes of the substructure network is carried out and they are renumbered from one to m, where m is the total number of internal nodes of the substructure. Boundary (main) nodes are assigned the subsequent numbers from $m + 1$ to n, where n is the total number of nodes in the substructure under consideration.

Transition of the form of eqn. (1.1) to the form (1.2) with renumbering of the knowns can be done using a special conversion matrix obtained from a common matrix of the same order as the stiffness matrix $[K]$ in eqn. (1.1). The conversion matrix is constructed by rearranging the coefficients in each column of a unit matrix into another row corresponding to the new number of the unknown.

For example, let the following be the structure of matrix $[K]$:

$$[\tilde{K}] = \begin{bmatrix} K_{11} & K_{12} & K_{13} & K_{14} \\ K_{21} & K_{22} & K_{23} & K_{24} \\ K_{31} & K_{32} & K_{34} & K_{34} \\ K_{41} & K_{42} & K_{43} & K_{44} \end{bmatrix}. \qquad \dots (1.8)$$

Let us assume q_1 and q_3 as boundary and q_2 and q_4 as internal unknowns and carry out their renumbering:

$$q_1 = q_3, \; q_3 = q_4, \; q_2 = q_1, \; q_4 = q_2$$

Conforming to the above renumbering, the unit matrix I_4 of the fourth order is converted into the following form:

$$[\tilde{I}_4] = \begin{bmatrix} 0 & 1 & 0 & 0 \\ 0 & 0 & 0 & 1 \\ 1 & 0 & 0 & 0 \\ 0 & 0 & 1 & 0 \end{bmatrix}. \qquad \dots (1.9)$$

It can readily be seen that the operation $[\tilde{I}][K][\tilde{I}]^T$ yields the matrix

$$[\tilde{K}] = \begin{bmatrix} K_{22} & K_{24} & K_{21} & K_{23} \\ K_{42} & K_{44} & K_{41} & K_{43} \\ K_{12} & K_{14} & K_{11} & K_{13} \\ K_{32} & K_{34} & K_{31} & K_{33} \end{bmatrix}, \qquad \dots (1.10)$$

i.e., yields the required structure of stiffness matrix blocks.

1.3.2 *The Second Variant of Condensation is Aimed at Maintaining the Original Banded Form of the Stiffness Matrix [K] in Eqn. (1.1) Obtained by Optimum Numbering of the Nodes. It Does Not Require the Arrangement of Unknowns in a Definite Sequence for Conversion into Form (1.2). In Principle, Unit Displacements Taken in Turn at Each of the Degrees of Freedom Provide in a Sequence at the Substructuring Boundary Nodes the Stiffness Coefficients Needed Even When all the Remaining Boundary Nodes are Fixed and Internal Nodal Displacements Occur (Meisner, 1968).*

The displacements found generate a conversion or transition matrix $[S]$ which establishes the relation between the boundary and internal nodal displacements:

$$\{q_i\} = [S]\{q_e\}.$$ ... (1.11)

By comparing eqn. (1.11) with the dependence between $\{q_i\}$ and $\{q_e\}$ as emerging from the second equation of system (1.2) at $P = 0$, i.e., without load on the internal nodes, we find:

$$[S] = \left[-K_{ii}^{-1} K_{ie}\right].$$ ... (1.12)

Considering that in the present case $K_{ie} = K_{ei}^T$, which can be verified in a physical sense, eqns. (1.5) and (1.6) can be represented as follows using eqn. (1.12):

$$\left[K_{ee}^*\right] = [K_{ee} + K_{ei} S],$$ ... (1.13)

$$\{P_e^*\} = \{P_e + S^T P_i\}.$$ ... (1.14)

1.4 Special Cases of Forming Stiffness Matrices

1.4.1 *Real Structures as a Rule Possess Symmetry and Periodicity, i.e., have Repetitive Members or Blocks in their Composition. When Designing Such Structures, the Use of Periodicity and Symmetry in the Course of Constructing Stiffness Matrices and Nodal Load Matrices for Different Levels of Superelements can Considerably Economise Calculations.*

This is particularly striking in the so-called 'unidimensional' scheme of synthesising substructures (Fig. 1.1) which accelerates the process of constructing the superelements by doubling the site of the superelement, i.e., splicing of two identical members of the preceding level (Shaposhnikov and Yudin, 1982, pp. 479–481).

Figure 1.1 shows how SE III of 8 SE lengths of zero level was constructed by doubling thrice. This process of substructuring is used, for example, in designing frame and high panelled buildings comprising repetitive members.

In some cases, it would be more convenient and effective to construct not enlarging but sliding superelements (Sapozhnikov, 1980a, b).

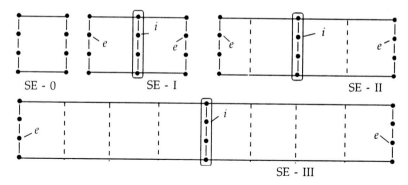

Fig. 1.1

Figure 1.2, a shows a unidimensional structure comprising some interconnected zero level SEs. By coupling two of the latter (Fig. 1.2, b) into one SE (Fig. 1.2, c) and eliminating the unknowns on the upper boundary, we produce a first level SE of the same length as that of zero level SE. By repeating the operation of eliminating the upper boundary nodes of the SE synthesised (Fig. 1.2, d, e, f) during subsequent build-up of the SE, we obtain an SE of the next level (SE IV in Fig. 1.2, f) having the same length as all the preceding level SEs.

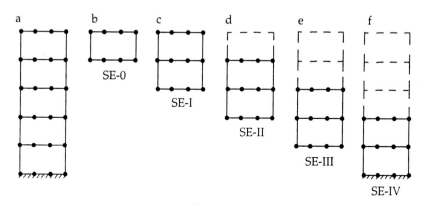

Fig. 1.2

An advantage of this process over expansion is the excellent manoeuvrability of the resultant system of equation. This is explained by the high coherence of the unknowns, i.e., by the high value of side-line secondary reactions r_{st} compared to the main reactions r_{tt} in the overall structural stiffness matrix.

1.4.2 *Considering that Excluding a Part of the Boundary Nodes When Forming the Stiffness Matrix of High Level SEs Using the General Procedure Described in Para 1.3.1 Disturbs the Displacements of Adjoining SEs in These Nodes, an Energy Variant of the Spline Interpolation of Displacements between the attachment-nodes, joining-nodes or under consideration is Often Used for Condensing the Stiffness Matrix (Voronenok and Sochinskii, 1981; Ignatiev and Kaurova, 1988; Shaposhnikov and Yudin, 1982, 1983; Araldsen and Roren, 1970).*

By equating the potential energy of the SE (Fig. 1.3) with the total stiffness matrix

$$PE = \frac{1}{2}\{q\}^T [K]\{q\}, \qquad \ldots (1.15)$$

the potential deformation energy of the SE with a reduced number of nodes and reduced stiffness matrix,

$$PE = \frac{1}{2}\{q_e\}^T [K_e]\{q_e\} \qquad \ldots (1.16)$$

and taking into consideration eqn. (1.11), we find

$$\{K_e^*\} = \left[K_{ee} + S^T K_{ii} S + K_{ei} S + (K_{ei} S)^T \right], \qquad \ldots (1.17)$$

In applying the nodal load on the SE to the nodes under consideration, the work of external forces of initial and reduced SEs is equated:

$$\{q_e\}^T \{P_e\} = \{q\}^T \{P\}. \qquad \ldots (1.18)$$

By substituting in eqn. (1.18) the dependence (1.11), we find

$$\{P_e^*\} = \{P_e + S^T P_i\} \qquad \ldots (1.19)$$

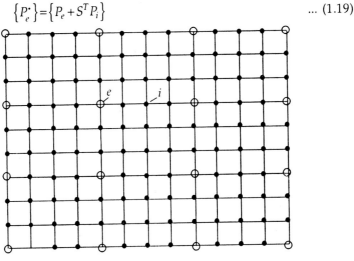

o Nodes considered (main)

• Nodes excluded (secondary or slave)

Fig. 1.3

Applying this condensation method in the initial computation of substructuring analysis is particularly effective when spline interpolation is used for nodal displacements in a periodic network of nodes in the form of lines dividing the structure into layers.

When an SE with a periodic or nearly periodic network of nodes is formed as a result of initial computation or forward sweep (Fig. 1.4), the backward sweep can be considerably accelerated without perceptibly sacrificing accuracy by using spline interpolation of displacements between the nodes under consideration (Ignatiev and Gorelov, 1986; Ignatiev, 1987).

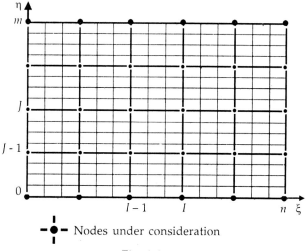

— ●— Nodes under consideration

Fig. 1.4

The displacements in the intermediate SE nodes can be expressed as displacements of nodes under consideration, forming a sparse network using the dependence

$$\{q_i\}=[S]\{q_e\}. \qquad \qquad \dots (1.20)$$

The approximating matrix $[S]$ is most simply constructed by spline interpolation. To form each of its lines corresponding to the line running parallel to axis ξ (or η) through node J (or I) in Figure 1.4, piece-polynomial spline interpolation in intermediate nodes I-1, I (or J-1, J) is carried out. Experience of spline interpolation shows that cubic splines are most effective.

To determine the internal nodal displacements not falling on the lines of the dispersed network of nodes, a second interpolation can be carried out. For it, the displacements determined in the first interpolation are used.

This algorithm is discussed in greater detail in Chapter 2.

References

Araldsen, P.O. March 1972. The application of superelement method in analysis and design of ship structures and machinery components. *National Symposium on Computerised Structural Analysis and Design*, Norway, pp. 17–39.

Araldsen, P.O. and E.M.Q. Roren. September 1970. The finite element method using superelements. *The SES M-69 Struct. Conference on Modern Techniques of Ship Struct. Analysis and Design*. University of California. Berkley.

Benfield, W.A. and R.F. Hruda. 1971. Vibration analysis of structures by component mode substitution. *AIAA J.*, Vol. 9, no. 7.

*Dashevskii, E.M. and M.V. Rotter. 1984. Application of substructuring method to solving problems of failure mechanics with a growing crack. *Problemy Prochnosti*, 7: 12–17.

*Gallagher, R. 1984. *Finite Elements Method*. Mir, 428 pp.

Hurty, W.C. 1965. Dynamic analysis of structural systems using component modes. *AIAA J.*, 3.

*Ignatiev, V.A. and S.F. Gorelov. 1986. Cellular system analysis by substructuring method with spline interpolation of transfers. *Izv. Vuzov. Stroitel'stvo i Arkhitektura*, 11: 30–34.

*Ignatiev, V.A. and T.M. Kaurova. 1988. *Developing a Dispersed Network Method*. Volgogr. Inzh.-Stroit. In-t., Volgograd, 16 pp. Deposited in VINITI, April 12, 1988, No. 2751-B88.

*Ignatiev, V.A. 1987. Berechnung von Kastenträger-systemen unter Verwendung von Spline-Superelement. *Technische Mechanik*, 8(4): 46–51.

*Ignatiev, V.A. and O.M. Ignatjewa. 1988. Untersuchung von Schwingungssystemen mit Hilfe eines frequenzdynamischen Verfahrens. *Wissenschaftliche Beiträge*, Technische Hochschule Wismar, Heft. Nov. 5, pp. 101–104.

*Ivantiev, V.I. and V.D. Chuban'. 1984. Analysing the forms and frequencies of free oscillations by multilevel dynamic condensation method. *Uchen. Zap. TsAGI*, 15(4): 81–92.

*Meisner, K. 1968. Multiconnected coupling algorithm for the method of structural analysis of stiffness. *Raketnaya Tekhnika i Kosmonavtika*, 11: 121–129.

*Postnov, V.A. and I.J. Kharkhuzim, 1974. Substructuring method for analysing the strength of ship structures. *Sudostroenie*, 11: 21–28.

*Postnov, V.A., S.A. Dmitriev, B.K. Eltyshev and A.A. Rodnionov. 1979. Substructuring method for analysing engineering structures. *Sudostroenie*, 288 pp.

*Przemieniecki, E.S. 1963. Matrix method of studying structures by substructure analysis. *Raketnaya Tekhnika i Kosmonavtika*, 1: 88–95.

*Przemieniecki, J.S. et. al.: Proceedings of Conference on "Matrix Methods in Structural Mechanics", Wright-Patterson Air Force Base, Ohio, 1965.

*Przemieniecki, J.S.: "Theory of Matrix Structural Analysis", McGraw Hill, 1968.

*Sapozhnikov, A.I. 1980a. Contour points method in structural analysis. *Stroitel'naya Mekhanika i Raschet Sooruzhenii*, 5: 59–61.

*Sapozhnikov, A.I. 1980b. Substructuring method in statics and dynamics of panelled buildings. *Izv. Vuzov. Stroitel'stvo i Arkhitektura*, 9: 33–37.

*Shaposhnikov, N.N. and V.V. Yudin. 1982. Application of the Method of Successive Doubling of Superelement for Determining the Frequency and Forms of Oscillations of Axisymmetrical Structures. MIIT, Moscow, 37 pp. Deposited in VINITI, October 4, 1983, No. 4658–B83.

*Shaposhnikov, N.N. and V.V. Yudin. 1983. Application of the Method of Successive Doubling of Superelements for Analysing Axisymmetrical Structures. MIIT, Moscow, 22 pp. Deposited in VINITI, October 4, 1983, No. 4659–B83.

*Smirnov, A.F., A.V. Aleksandrov, B.Ya. Lashchenikov and N.N. Shaposhnikov. 1981. *Structural Engineering (Pivotal Systems)*. Stroiizdat, Moscow, 512 pp.

*Voronenok, E.Ya. and S.V. Sochinskii. 1981. Interpolation condensation of stiffness matrix when solving structural engineering problems by substructuring method. *Prikladnaya Mekhanika*, 17(67): 114–118.

Wilson, E.L. 1974. The static condensation algorithm. *Int. J. Num. Math. Engg.* 8: 198–203.

---

*Asterisked entries are in Russian—General Editor.

2

Design of Thin-walled Plate and Box-type (Cellular) Systems by the Substructuring Method

2.1 Introduction

The spatial thin-walled cellular systems examined in this chapter are comprised of flat plates. The plate members may vary in thickness, apertures and notches inside and on the edge. It has been assumed that the plate material of the cellular system functions as a linear-elastic body, i.e., obeys Hooke's law. The plate thickness is regarded as much less than its other dimensions and, as a result, plate displacements and deformations of given structures are determined by the corresponding displacements and deformations of their middle surface. Each plate of the box-type (cellular) system can then be divided into planar finite elements (FE), each node of which generally has six degrees of freedom (three displacements along the co-ordinate axes and three rotation angles around these axes).

2.2 Reduced Stiffness Matrix of a Box-type Structure for Substructuring Analysis

2.2.1 Some Aspects of Formation of the Stiffness Matrix (SM) of a Cellular System by the Finite Element Method

Let us derive the system of FEM equations in the form of the displacement method for cellular systems. The main system is obtained by introducing at all the nodes of the FE network kinematic degrees of freedom in the six directions of linear and angular displacements: x, y, z, φ^x, φ^y, φ^z.

Depending on the type of cellular construction, it can be represented by triangular or rectangular finite elements or their combinations. Thus for cellular shells comprising oblique angled plates, triangular finite elements with an additional node on the oblique edge increase the analytical accuracy and are advantageous (*Maslennikov, 1987).

In a general case, the nodal displacement vector of a rectangular FE is as follows:

$$\{q_j\} = \begin{bmatrix} u_j & v_j & w_j & \varphi_j^x & \varphi_j^y & \varphi_j^z \end{bmatrix}^T, \qquad \text{... (2.1)}$$

The SM transformation matrix of an FE from a local into a global coordinate system is quasidiagonal (Rice, 1984):

$$\begin{bmatrix} \Lambda^{(r)} \end{bmatrix} = \begin{bmatrix} \begin{bmatrix} \tilde{\lambda}^{(r)} \end{bmatrix} & & & \\ & \begin{bmatrix} \tilde{\lambda}^{(r)} \end{bmatrix} & & \\ & & \begin{bmatrix} \tilde{\lambda}^{(r)} \end{bmatrix} & \\ & & & \begin{bmatrix} \tilde{\lambda}^{(r)} \end{bmatrix} \end{bmatrix}, \qquad \text{... (2.2)}$$

where

$$\begin{bmatrix} \tilde{\lambda}^{(r)} \end{bmatrix} = \begin{bmatrix} \begin{bmatrix} \lambda^{(r)} \end{bmatrix} & \\ & \begin{bmatrix} \lambda^{(r)} \end{bmatrix} \end{bmatrix}, \qquad \text{... (2.3)}$$

Here $\begin{bmatrix} \lambda^{(r)} \end{bmatrix}$ represents the cosine directrix matrix of the local co-ordinate system. In a general case, this matrix is as follows:

$$\begin{bmatrix} \lambda^{(r)} \end{bmatrix} = \begin{bmatrix} l_1 & l_2 & l_3 \\ m_1 & m_2 & m_3 \\ n_1 & n_2 & n_3 \end{bmatrix}. \qquad \text{... (2.4)}$$

The first column of this matrix contains the cosine directrix of axis $ox^{(r)}$, the second column of axis $oy^{(r)}$ and the third of axis $oz^{(r)}$.

When ignoring shear deformations, it should be remembered that, for the nodes of an FE mesh in which coplanar elements are coupled, the equilibrium equations corresponding to the rotation around the axis perpendicular to the plane of these elements are transformed into the identity O=O. Since, in this case, zeros appear on the main diagonal of the SM of the cellular system with orthogonal walls, it becomes a degenerate SM. To resolve the corresponding FEM algebraic system of equations, the rows and columns corresponding to the zero-order reactions should be eliminated.

Analysis of a membrane cellular system whose walls take the load only in its plane is similar. The nodal displacement vector in this case is as follows:

$$\{q\} = \begin{bmatrix} u_j & v_j & w_j \end{bmatrix}. \qquad \text{... (2.5)}$$

Since displacements w_j from the plane of a membrane FE do not influence the stress-strain state in its plane, zero rows and columns corresponding to displacement w_j appear in the SM of this element. These rows and columns corresponding to the zero-order reactions in the nodes should be eliminated from the general SM.

The FEM system of equations in the displacement method are as follows:

$$[K]\{q\} + \{P\} = 0, \qquad \qquad \text{... (2.6)}$$

where $[K]$, $\{q\}$ and $\{P\}$ are the SM of the system, vector of joint displacements and vector of nodal loadings respectively. Kholetskii's (in Shoop, 1982) resolution of the SM $[K]$ using the square root method is depicted in the form of the product of the lower by the upper triangular matrix:

$$[K] = [L][L]^T. \qquad \qquad \text{... (2.7)}$$

By substituting the above equation in (2.6) and introducing new variables

$$\{y\} = [L]^T\{q\}, \qquad \qquad \text{... (2.8)}$$

the following system of equations is resolved

$$[L]\{y\} = \{P\}, \qquad \qquad \text{... (2.9)}$$

and later the displacements $\{q\}$ are found from eqn. (2.8).

The system of eqns. (2.8) and (2.9) have triangular matrices of coefficients of unknowns and hence their solution poses no problems.

2.2.2 Some Aspects of Formations of the Stiffness Matrix of a Cellular System by the Substructuring Method

Unlike the general substructuring method, for each plate of the cellular system let us have, as suggested in (Segerlind, 1979; Sapozhnikov and Gorelov, 1982; *Finite Element Method in the Mechanics of Solid Bodies*, 1982; Novitskii, 1984) an enlarged network of nodes (points under consideration) and the corresponding basic degrees of freedom to which the loading and stiffness parameters of the cellular system are assigned. The remaining nodes (additional degrees of freedom) of the FE are eliminated by a single or multiple level substructuring procedure. The points under consideration are selected on the basis of the following characteristics of the cellular system under consideration:

(1) the gradient of the effective load (the gentler the variation of load value and direction along the length or width of the plate, the more widely located the points);

(2) presence of notches in the plate and abrupt changes of stiffness parameters along the length or width (points must be located along the edges of notches and on the lines of abrupt stiffness variation); and

(3) design accuracy required.

Let us represent the initial eqn. (2.6) in the following form:

$$\begin{bmatrix} K_{rr} & K_{rs} \\ K_{sr} & K_{ss} \end{bmatrix} \begin{Bmatrix} q_r \\ q_s \end{Bmatrix} + \begin{Bmatrix} P_r \\ P_s \end{Bmatrix} = 0. \qquad \ldots (2.10)$$

Here, subscript 'r' refers to the basic degrees of freedom, which are the retained co-ordinates, and 's' to the additional degrees, which are the unwanted co-ordinates to be eliminated later.

From the second eqn. of (2.10), we find

$$q_s = - K_{ss}^{-1} K_{sr} q_r - K_{ss}^{-1} P_s. \qquad \ldots (2.11)$$

By substituting this result in the first eqn. of (2.10), we obtain a reduced system of equations

$$\left[\tilde{K}_{rr} \right] \{q_r\} + \{\tilde{P}_r\} = 0, \qquad \ldots (2.12)$$

where

$$\left[\tilde{K}_{rr} \right] = \left[K_{rr} - K_{rs} K_{ss}^{-1} K_{sr} \right],$$

$$\{\tilde{P}_r\} = \left\{ P_r - K_{rs} K_{ss}^{-1} P_s \right\} \qquad \ldots (2.13)$$

are respectively the SM and nodal loading matrix reduced to the points under consideration.

Eqns. (2.13) are known as the equations of static condensation of the substructuring method (*Finite Element Method in the Mechanics of Solid Bodies*, 1982; Sapozhnikov and Gorelov, 1982; and others). Kholetskii's resolution is best used to simplify the application of matrix K_{ss}. For Kholetskii's resolution the square roots method is generally used. It is the quickest and one of the most effective methods on a computer.

2.2.3 Reduction of Stiffness Matrix in Substructuring Analysis Using a Transformation Matrix

Let us examine the approach in which reduction (condensation) is based on the concept of a transformation matrix. The problem then involves constructing the relationship

$$\{q_s\} = [S] \{q_r\}, \qquad \ldots (2.14)$$

where [S] is the unknown conversion matrix called the transformation matrix.

Since the second member in eqn. (2.11) is a constant for a given loading and corresponds, in a physical sense, to the reactions caused by loading P_s in the nodes in the direction of additional degrees of freedom in the absence of displacements in the directions of the main degrees of freedom, the ratios between the degrees of freedom q_s and q_r are fixed using the matrix

$$[S] = - K_{ss}^{-1} K_{sr}. \qquad \ldots (2.15)$$

14

Considering that in this case $K_{rs} = K_{sr}^T$ in a physical sense, using eqn. (2.15), eqns. (2.13) can be represented in the following form:

$$\left[\tilde{K}_{rr}\right] = \left[K_{rr} - K_{rs}\,S\right], \qquad\qquad \text{... (2.16)}$$

$$\left[\tilde{P}_r\right] = \left[P_r + S^T\,P_s\right].$$

From the physical concept of the problem, the columns of transition matrix $[S]$ in eqn. (2.14) represent the internal node displacements superelements (with additional degrees of freedom) due to successive singular shift of nodes located on the boundary.

Let us examine in greater detail the structure of matrix $[S]$ on the example of a plate superelement. Figure 2.1 shows the plate superelement and the zone of non-zero deflections (hatched portion) on a unit displacement from the plane of a superelement of the i-th boundary node and the other nodes fixed on the boundary.

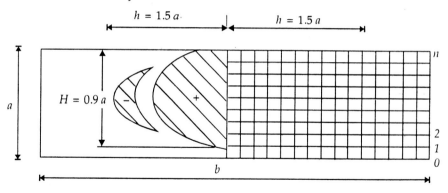

Fig. 2.1

The internal nodes of the FE network are free.

FEM calculations established that, for a superelement in the form of a rectangular plate fixed along the contour or a regular system of crossbeams of width a and arbitrary length $b > 3a$ (in the calculations, $b = 5a$), unit displacement of the i-th boundary node on the long side of the SE contour hardly causes shifts of internal nodes of the FE network (FE—plates or beams) at distance $h \geq 1.5a$ toward the left or right of the i-th node to 'depth' $H \geq 0.9a$.

Figure 2.2 shows the dependence of the ratio of internal node displacements on an arbitrary 'zero' boundary to a unit displacement of the i-th node on the SE contour on the number of rows n along the width of the plate or regular system of crossbeams.

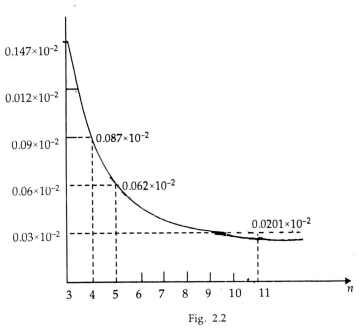

Fig. 2.2

During displacement of the i-th node in the other directions and unit rotations, the non-zero displacement zone of internal nodes is even less.

Figure 2.3 depicts the matrix $[S]$ structure. It shows that matrix $[S]$ is rectangular since the number of boundary displacements (basic degrees of freedom—r) is not equal to the number of internal displacements (additional degrees of freedom—s). Let us call this matrix a quasi-strip unlike the square matrix strip containing the main and secondary diagonals. The strip $[S]$ width depends on the plate width and the numeration order of the boundary and internal nodes.

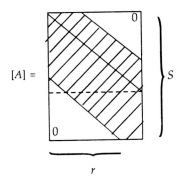

$$[A] =$$

Fig. 2.3

Completeness of matrix [S] depends on the ratio a/b. The closer this ratio is to 1, the greater will be the completeness of matrix [S].

The numerical experiment shows that a typical plate section with ratio of sides 1:3 and FE on the shorter side not exceeding 11 is best adopted as a superelement. By imparting unit generalised displacement to the i-th node on the superelement contour in all the axial directions, let us determine the displacements of internal nodes of the type superelement using the common FEM, assuming that all the remaining boundary nodes are fixed, using a full set of nodes. Let the i-th column of matrix [S] be filled with the displacement values of the internal nodes obtained. The two symmetrical axes of the SE help reduce to one-fourth the total number of arithmetical operations. Thus reduction results in forming only blocks K_{rr}, K_{rs} and non-zero members of matrix [S]. To construct matrix [S] allowing for SE symmetry, the FEM system of equations for type SE will have to be solved $2s$ times, where s is the total number of degrees of freedom of the nodes on the SE's shorter side.

The efficiency of the proposed method of direct construction of the transformation matrix [S] can be evaluated by comparing the number of arithmetical operations by this method versus other methods.

The initial computation of the stiffness matrix of internal nodes (additional degrees of freedom) $[K_{ss}]$ in eqn. (2.15) requires a large number of operations and for reasons of practically is not adopted in the substructuring method. One of the most effective methods for solving the linear algebraic system of equations (LASE) is the LU resolution (Rice, 1984, p. 52) of the matrix of its coefficients. This resolution requires some $n^3/3$ arithmetical operations, where n is the order of the LASE (in our case, n is the number of reduced internal degrees of freedom of the SE).

In the direct construction of matrix [S], the system of equations for the suggested type SE has an order of $n = 3a^2$ with a strip width of non-zero coefficients $d = s$. It is known (Rice, 1984, p. 156) that the number of arithmetical operations when resolving LASE with a strip structure of coefficients is about $d^2 n$. Hence the total number of arithmetical operations with allowance for the symmetry of the system $N = S^2 \times 3 S^2 \times 2 S = 6 S^5$.

Thus, for the same order of the system of equations, the proposed method requires 1.5 S fewer arithmetical operations than the LU resolution.

The computing efficiency of the proposed algorithm is greater the larger the number of SE internal degrees of freedom s. In a general case, the economy in the number of operations is $k^2 s/6$, where $k = b/a$ for an SE finite element square cell network.

We have examined the transformation matrix [S] for a bar or ordinary plate. Let us now generalise this method for analysing a cellular system.

A fragment of the cellular system is divided into transverse sections such that the height of each part p is 3 to 4 times smaller than the other dimensions (a and b) along the contour (Fig. 2.4).

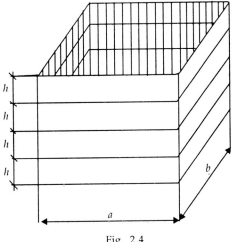

Fig. 2.4

Let us examine one part of the fragment of the cellular system as an independent SE. Let us divide this SE into FEs and classify their nodes into internal and boundary (Fig. 2.5) and connect these boundary nodes with external links.

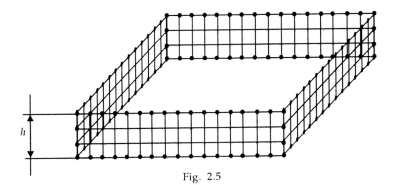

Fig. 2.5

It is quite evident that as one of the boundary nodes on any plate of the SE under consideration (for example, node i in the strip front) is given unit displacement, the shifts of internal nodes on the other plates of this SE will be equal to zero. The shifts of internal nodes on the front side of the plate will be, as shown before, of a local character. A similar picture is noticed during the shifts of boundary nodes on other sides.

Thus, any fragment of a cellular system can be divided into type superelements-plates for which transformation matrix [S] can be constructed by the proposed method after which stiffness matrix reduction of the SE and the loading of boundary nodes are carried out. By combining the SE of the first level within two transverse sections of the cellular system, a closed SE as shown in Fig. 2.5 is obtained.

When there are apertures in the cellular system, the division into fragments should be such that the openings fall between the lines of boundary points (Fig. 2.6).

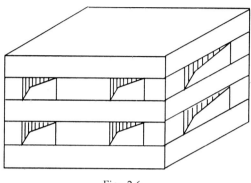

Fig. 2.6

For parts with openings, each partition should be examined and matrix [S] constructed from unit displacements of its boundary nodes. Clearly, such a procedure for identical openings can be done at one time.

Cellular systems consisting of plates of different thickness are regarded as similar to systems with apertures, i.e., are divided by lines with boundary nodes into parts (SEs) having the same thickness and an independent matrix [S] constructed for each part.

The following conclusion may be drawn although the figures depict the fragments of cellular systems only in the form of 'four walls', all discussions are applicable to any cellular system with mutually orthogonal arrangement of planar elements.

2.3 Two-dimensional Interpolation of a Displacement Field by Displacement Values at Selected Points

Let us assume that each plate or a tapered shell comprising the cellular structure under consideration is divided into finite elements (FE) whose boundaries form a two-dimensional orthogonal network. Adopting these FEs as SEs of the first level, let us mark on the initial network of these SEs a dispersed mesh formed by the boundaries of second level SEs. Let us adopt the nodes in this mesh as selected points for substructuring analysis (Sapozhnikov, 1980). The SE dimensions are defined by the gradient of the prevalent load and the need to allow for sudden changes in the plate stiffness parameter.

After determining the displacements at selected points, the cellular system is divided into individual plates and a subvector of the displacements of the dispersed mesh at selected points of a given plate f_p^i is isolated for each i-th plate of the general vector of displacement at selected points f_p. It must be pointed out that it is not the fragments of the plate falling in the SE during formation of the stiffness matrix of selected points that are taken into consideration, but the plates as a whole even if their constituent parts pertain to a few SEs. The SEs themselves have already been used during formation of the stiffness matrix of the selected points of the cellular system and are no longer considered. The plates may be randomly oriented relative to the co-ordinate axes; further, the isolated surfaces of the cellular system may be represented as tapered shells and not as plates. These surfaces affect only the selection of the basic FEs in the first level SEs (Norri, 1981) and are determined by the same method employed for plates and tapered shells forming the members of the cellular system in the stage of interpolation of displacements at internal nodes. It is assumed that the thickness of the isolated plate (shell) is constant and that the material is elastic.

2.3.1 Order of Two-dimensional Interpolation of Displacement Field by Displacement Values at Selected Points

Interpolation of the displacement field, especially in the FEM, can generally be carried out by different methods: Hermitian polynomials, Lagrange, polynomials of Hedmit Stirling, Bessel, Gauss, Yakobi, Gegenbauer, Chebyshev, trigonometric and others (Bate and Wilson, 1982; Aleksandrov et al., 1983; Zavylov, 1980; Segerlind, 1979; Norri, 1981; Smirnov et al., 1984; Zenkevich and Morgan, 1986). Each of these interpolation polynomial has its own advantages and deficiencies and interpolation giving an accurate solution depends on the type of the polynomial selected as well as on the number of interpolation nodes, parameters of the structure considered, loading type, fixing *interboundary conditions* and other factors. During the interpolation of a displacement field within an FE in a two-dimensional case, unidimensional functions of the following form are usually multiplied (Segerlind, 1979; Bate and Wilson, 1982; Aleksandrov et al., 1983; Norri, 1981):

$$\Delta = N_\xi \times R \times N_\eta, \qquad \qquad ... (2.17)$$

where Δ is one of the displacement vector components of an arbitrary point within the FE; N_ξ and N_η vectors of the form functions in the directions ξ and η; and R the matrix of generalised nodal displacements in the FE.

Thus, eqn. (2.17) yields an interpolation equation for a two-dimensional FE. However, when the nodes in each direction are more, this procedure of multiplying the unidimensional functions becomes cumbersome and ineffective (Aleksandrov et al., 1983; Bate and Wilson, 1976, 1982). Let us examine the first operation of multiplication in eqn. (2.17):

$$F_\xi = N_\xi \times R. \qquad \qquad \cdots (2.18)$$

Here, the matrix-line F_ξ has an inbuilt equation for unidimensional interpolation of internal displacements of the FE for each line of nodes in the FE in the axis direction ξ. This can be illustrated by a simple example.

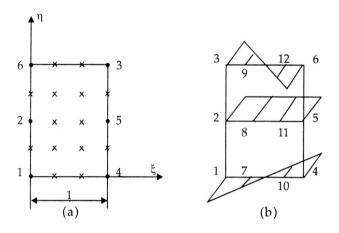

Fig. 2.7

The FE shown in Figure 2.7, a has three nodal lines (1–4, 2–5 and 3–6) parallel to axis ξ and two parallel to axis η. Matrix R and vector N_ξ have the following form:

$$R = \begin{bmatrix} \Delta_1 & \Delta_2 & \Delta_3 \\ \Delta_4 & \Delta_5 & \Delta_6 \end{bmatrix}, \; N_\xi = \left[(1 - \xi)\xi \right]. \qquad \cdots (2.19)$$

Substituting these values in eqn. (2.18), we derive

$$F = \left[\Delta_1(1-\xi)+\Delta_4\xi; \; \Delta_2(1-\xi)+\Delta_5\xi; \; \Delta_3(1-\xi)+\Delta_6\xi \right]. \qquad \cdots (2.20)$$

Every member of F represents an equation for unidimensional interpolation along ξ (Fig. 2.7,b). On multiplying F by vector N_η, the FE internal displacement values along lines 1–4, 2–5 and 3–6 do not change since the functions of forms along each direction are mutually independent. By this multiplication operation, we 'fix' the law of transition from shifts along line 1–4 to shifts along line 2–5 and later line 3–6 (Fig. 2.7,b). If the internal displacements are required only at some points of this FE (e.g., at points marked by crosses in Fig. 2.7,a), the displacements at the additional points 7–12 (Fig. 2.7,b) are initially determined by using the equation for F. The above matrix R^* is constructed as

$$R^* = \begin{bmatrix} \Delta_1 & \Delta_2 & \Delta_3 \\ \Delta_4 & \Delta_5 & \Delta_6 \\ \Delta_7 & \Delta_8 & \Delta_9 \\ \Delta_{10} & \Delta_{11} & \Delta_{12} \end{bmatrix}. \qquad \ldots (2.21)$$

Later, matrix R^* is multiplied by vector N_η, as a result of which equations are derived for interpolation along axis η. From the resultant equations, the rest of unknown displacements are defined.

This simple and fairly clear method is also used for the interpolation of displacements of the nodes of FEs network in a fragment of cellular system through displacements of enlarged network of selected (main) points ij (Fig. 2.8).

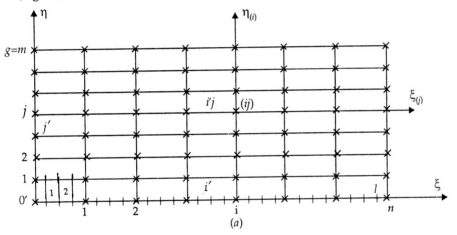

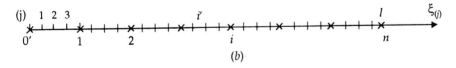

Fig. 2.8

Let the network of selected points lie at the initial network of FEs in the fragment under consideration (plate or shallow shell) in the local system of co-ordinates $\xi 0' \eta$. The network has $(m + 1)$ lines in the direction ξ and $(n + 1)$ lines in the direction η. On each j-th line along axis ξ are marked $t = n + 1$ selected points $(j = 0, \ldots, m)$ and on each i-th line along axis η $g = m + 1$ points $(i = 0, \ldots, n)$. Unidimensional interpolation is initially carried out in the direction of axis ξ on the lines of the mesh passing through point η_j $(j = 0, \ldots, m)$. Let the displacements of the nodes $i'j$ be interpolated by a polynomial of degree d:

$$P(\xi_i, \eta_j) = \alpha_0 + \alpha_1\xi_{i'} + \alpha_2\xi_{i'}^2 + \ldots + \alpha_d\,\xi_{i'j}^d. \qquad \ldots (2.22)$$

The following equals conditions should hold at the points under consideration on each line:

$$P(\xi_i, n_j) = (f_p)_{ij}, \qquad \begin{pmatrix} j=0,1,\ldots,m \\ i=1,2,\ldots,t \end{pmatrix}, \qquad \ldots (2.23)$$

where $(f_p)_{ij}$ is the displacement (along axes x, y or z) of the i-th selected point on the j-th line in the direction of axis ξ and p the conventional sign of the selected point.

Equal (2.23) give us the possibility to determine the interpolation coefficients in (2.22).

The displacements of internal nodes $i'j$ falling on the lines parallel to axis ξ and passing through the points of the mesh η_j ($j = 0, \ldots, m$) are determined by interpolation(2.22). Let there be nodes on the line passing through point $\xi_{j'}$ of which t are the selected nodes and $l - t$ internal. Interpolation is then continued along axis η on lines passing through points ξ_i ($i' = 0,1, \ldots, l$) using a polynomial analogous to (2.22):

$$P(\xi_{i'}, \eta_{j'}) = \tilde{\alpha}_0 + \tilde{\alpha}_1\eta + \tilde{\alpha}_2\,\eta_{j'}^2 + \ldots + \tilde{\alpha}_d\,\eta_{j'}^d, \qquad \ldots (2.24)$$

considering the equation

$$P(\xi_{i'}, \eta_j) = (f_p)_{i'j} \qquad (j=0,1,2,\ldots,m), \qquad \ldots (2.25)$$

where $(f_p)_{i'j}$ is the displacement (along axes x, y or z) of the i'-th joint on the j-th line along the axis ξ [it must be pointed out that these displacements pertain not only to the selected points but also internal nodes for which shifts were determined by interpolation using eqns. (2.22) and (2.23)].

The using of eqns. (2.25) let us determine the interpolation coefficients $\tilde{L}$ and with (2.24) the displacements $(f)_{i'j'} = P(\xi_{i'}, \eta_{j'})$.

By following the above procedure for each surface of the cellular system, displacements are determined for all the internal nodes, by passing the phased backward sweep of substructuring.

2.4 Methodology of Determining Displacements in the Substructuring Analysis of Plate (Cellular) Systems Using Splines

2.4.1 Use of Interpolation Polynomials of Lagrange and Hermite

Let us examine again the fragment of the cellular system in the local co-ordinate system $\eta 0'\xi$ (Fig. 2.8) with a network of selected points having $(m + 1)$ lines in the direction of axis ξ and $(n + 1)$ lines in the direction of axis

η. Since the number of nodes on each line of the dispersed network may differ, the use of splines is most convenient in the interpolation procedure.

Let us study the spline construction based on Lagrange interpolation coefficients. The interpolation functions in the form of Lagrange polynomials of the n-th order are determined from the following well-known equation of Berezin and Zhidkov (1966):

$$L_i^n = \frac{(x-x_1)(x-x_2)....(x-x_{i-1})(x-x_{i+1})....(x-x_n)}{(x_i-x_1)(x_i-x_2)..(x_i-x_{i-1})(x_i-x_{i+1})..(x_i-x_n)}, \qquad \text{... (2.26)}$$

where x is the current co-ordinate and $x_1,...,x_n$ the co-ordinates of interpolation nodes.

The order of the interpolating polynomial should not exceed that of the differential equation defining the structural deformation under load (Ignatiev, 1973; Ignatieva, 1983; Zivograd, Dragolub, 1997; Tongchen et al., 1996, Krjukov, 1997). When studying unidimensional interpolation with no load between the selected points, Lagrange polynomials can be restricted to the third degree. When the load q is uniformly distributed between the selected points, polynomials of the fourth degree are used; when the load distribution in the interval between the selected points conforms to the rule $(qx + b)$, a fifth degree polynomial is used. For shallow isotropic shells, similar restrictions are worked out when selecting the interpolating polynomial.

The methodology of spline interpolation is examined in detail below on the example of using Lagrange third degree polynomials. The method is similar for polynomials of other degrees.

Let us examine a section of the j-th line along axis ξ of the network of selected points from the $(i-1)$-th through $(i + 2)$-th point (Fig. 2.9) and formulate the function of the form of the Lagrange polynomial.

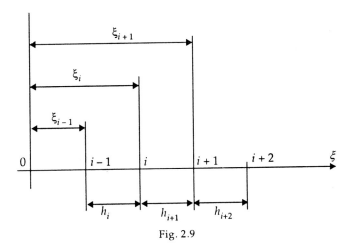

Fig. 2.9

The forms for the interpolation functions for this special case are found for $\xi_{i-1} \geq \xi \geq \xi_{i+2}$ using eqn. (2.26):

$$L_{i-1}^{\xi} = \frac{(\xi - \xi_i)(\xi - \xi_{i+1})(\xi - \xi_{i+2})}{-h_i(h_i + h_{i+1})(h_i + h_{i+1} + h_{i+2})},$$

$$L_i^{\xi} = \frac{(\xi - \xi_{i-1})(\xi - \xi_{i+1})(\xi - \xi_{i+2})}{h_i h_{i+1}(h_{i+1} + h_{i+2})},$$

$$L_{i+1}^{\xi} = \frac{(\xi - \xi_{i-1})(\xi - \xi_i)(\xi - \xi_{i+2})}{-h_{i+1}(h_{i+1} + h_i)h_{i+2}}, \qquad \ldots (2.27)$$

$$L_{i+2}^{\xi} = \frac{(\xi - \xi_{i-1})(\xi - \xi_i)(\xi - \xi_{i+1})}{(h_i + h_{i+1} + h_{i+2})(h_{i+1} + h_{i+2})h_{i+2}}.$$

Interpolation is first done on the lines of the network passing through point $\eta_j (j = 0, ..., m)$. A spline is constructed on each line for $\xi \in [\xi_{i-1}, \xi_{i+2}]$ using the following equations:

$$S^i(\xi, \eta_j) = L_{i-1}^{\xi}(f_p)_{i-1,j} + L_i^{\xi}(f_p)_{i,j} + L_{i+1}^{\xi}('f_p)_{i+1,j} + L_{i+2}^{\xi}(f_p)_{i+2,j}, \qquad \ldots (2.28)$$

$$(j = 0, ... m; \ i = 1, ... t)$$

where L_i^{ξ} are the Lagrange interpolation coefficients from eqn. (2.27); and $(f_p)_{i,j}$ the displacement (along axes x, y or z) of the j-th line in the direction ξ.

The displacements of internal nodes in the section from ξ_{i-1} to ξ_{i+2} are determined from spline (2.28). By substituting the spline in one section, we find from eqn. (2.27), with appropriate substitution of indices, the interpolation coefficients L^{ξ} for the section $[\xi_i, \xi_{i+3}]$. By substituting the coefficients in eqn. (2.28), we construct spline $S^{i+1}(\xi, \eta_j)$ (upper index of the spline denotes the number of the first interval between the selected points figuring in a given spline). Having focussed on the concrete section h_{i+1} between points i and $i + 1$, we find that the displacements within the section are interpolated thrice since it figures in splines $S^{i-1}(\xi, \eta_j)$, $S^i(\xi, \eta_j)$ and $S^{i+1}(\xi, \eta_j)$ (Fig. 2.10).

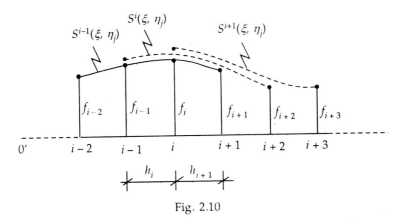

Fig. 2.10

This greatly improves the accuracy of the displacement values of internal nodes. They are defined as the average of three values obtained from three adjoining splines:

$$\left(f_p\right)_{s,j} = \frac{1}{3}\left[\left(f_p\right)_{s,j}^{i-1} + \left(f_p\right)_{s,j}^{i} + \left(f_p\right)_{s,j}^{i+1}\right], \qquad \ldots (2.29)$$

where s is the index of any point in the section h_{i+1}, i.e., in the interval $[i, i+1]$.

To determine the rotation angles of internal nodes, the derivatives of the interpolation functions of the following forms are first determined:

$$\theta_{i-1}^{\xi} = \frac{dL_{i-1}^{\xi}}{d\xi} = \frac{3\xi^2 - 2\xi\left(\xi_i + \xi_{i+1} + \xi_{i+2}\right) + \xi_i\left(\xi_{i+1} + \xi_{i+2}\right) + \xi_{i+1}\cdot\xi_{i+2}}{-h_i\left(h_i + h_{i+1}\right)\left(h_i + h_{i+1} + h_{i+2}\right)},$$

$$\theta_i^{\xi} = \frac{dL_i^{\xi}}{d\xi} = \frac{3\xi^2 - 2\xi\left(\xi_{i-1} + \xi_{i+1} + \xi_{i+2}\right) + \xi_{i-1}\left(\xi_{i+1} + \xi_{i+2}\right) + \xi_{i+1}\cdot\xi_{i+2}}{-h_i + h_{i+1}\left(h_i + h_{i+1} + h_{i+2}\right)},$$

$$\ldots (2.30)$$

$$\theta_{i+1}^{\xi} = \frac{dL_{i+1}^{\xi}}{d\xi} = \frac{3\xi^2 - 2\xi\left(\xi_{i-1} + \xi_i + \xi_{i+2}\right) + \xi_{i-1}\left(\xi_i + \xi_{i+2}\right) + \xi_i\cdot\xi_{i+2}}{-\left(h_i + h_{i+1}\right)h_{i+1} + h_{i+2}},$$

$$\theta_{i+2}^{\xi} = \frac{dL_{i+2}^{\xi}}{d\xi} = \frac{3\xi^2 - 2\xi\left(\xi_{i-1} + \xi_i + \xi_{i+1}\right) + \xi_{i-1}\left(\xi_i + \xi_{i+1}\right) + \xi_i\cdot\xi_{i+1}}{\left(h_i + h_{i+1} + h_{i+2}\right)\left(h_{i+1} + h_{i+2}\right)h_{i+2}}.$$

The rotation angles of internal nodes falling on the mesh lines along the direction of axis ξ and passing through point $\eta_j(j=0,\ldots,m)$ are determined using the following equation:

$$\varphi^i(\xi,\eta_j)=\theta_{i-1}^\xi(f_p)_{i-1,j}+\theta_i^\xi(f_p)_{i,j}+\theta_{i+1}^\xi(f_p)_{i+1,j}+\theta_{i+2}^\xi(f_p)_{i+2,j}+$$

$$\Delta(\xi,\eta_j), \qquad\qquad\qquad \text{... (2.31)}$$

where θ_i^ξ is the function from eqn. (2.30); $(f_p)_{i,j}$ is the displacement of the i-th estimated point on the j-th line; $\Delta(\xi,\eta_j)$ is an additional member necessitated by the original lack of coincidence; (difference) in the co-ordinate systems xoy and $\xi0'\eta$ (Fig 2.8).

If the co-ordinate systems xoy and $\xi0'\eta$ coincide or parallel (a plate in the plane xoy from a cellular system is under consideration), $\Delta(\xi_i\eta_j)=0$. When an ordinary plate is under consideration and the systems xoy and $\xi0'\eta$ do not coincide, the additional member $\Delta(\xi,\eta_j)$ is determined only once from the known geometric proportions and it remains constant for all nodes in the plate under consideration.

In the selected points of the cellular system falling on the lines joining surfaces running in different directions, not only the displacements but also the rotation angles around the co-ordinate axes are determined. This helps fix the boundary conditions of displacements and rotation angles in the selected points falling on the contour of the surface under consideration when examining the fragments of a cellular system. For rotation angles along the surface contour in the sections adjoining the contour, Hermite's displacement interpolation for $\xi\in[\xi_{i-1},\xi_i]$ by $i=1$ or $i=n$, is carried out:

$$S^i(\xi,\eta_j)=F_1^\xi(f_p)_{i-1,j}+F_2^\xi(\varphi_p)_{i-1,j}+F_3^\xi(f_p)_{i,j}+F_4^\xi(\varphi_p)_{i,j}, \qquad \text{... (2.32)}$$

where F_s^ξ are the Hermite functions:

$$F_1^\xi=1-3r^2+2r^3; \qquad F_2^\xi=\xi(1-2r+r^2);$$

$$F_3^\xi=3r^2-2r^3; \qquad F_4^\xi=\xi(r^2-r);$$

$$r=\xi/h_i,$$

$(\varphi_p)_{i,j}$ are the rotation angles of the section in the selected points on the contour known from the boundary conditions and converted into the local co-ordinate system $\xi0'\eta$ or obtained from eqn. (2.31) without the additional member $\Delta(\xi,\eta_j)$ (for nodes not falling on the contour but adjoining it).

The rotation angles of internal nodes are determined from the following interpolation equation:

$$\varphi^i\left(\xi,\eta_j\right)=\psi_1^\xi\left(f_p\right)_{i-1,j}+\psi_2^\xi\left(\varphi_p\right)_{i-1,j}+\psi_3^\xi\left(f_p\right)_{i,j}+\psi_4^\xi\left(\varphi_p\right)_{i,j}+\Delta\left(\xi,\eta_j\right),$$

$$\ldots (2.33)$$

where φ_s^ξ is determined from the equation $\psi=\partial F/\partial\xi$:

$$\psi_1^\xi=6\left(r^2-r\right)/h; \qquad \psi_2^\xi=\left(1-4r+3r^2\right);$$

$$\psi_3^\xi=6\left(r^2-r\right)/h; \qquad \psi_4^\xi=\left(3r^2-2r\right). \qquad \ldots (2.34)$$

Let there be l nodes on each line in the direction ξ (Fig. 2.8): t represents the selected and $l-t$ internal (for all lines passing through point $\eta_j(j=0,1,..,m)$. A spline is constructed on each line in direction η for the interval $\eta\in\left[\eta_{j-1},\eta_{j+2}\right]$ as follows:

$$S^j\left(\xi_i,\eta\right)=L_{j-1}^\eta\left(f_p\right)_{i,j-1}+L_j^\eta\left(f_p\right)_{i,j}+L_{j+1}^\eta\left(f_p\right)_{i,j+1}+L_{j+2}^\eta\left(f_p\right)_{i,j+2}, \quad \ldots (2.35)$$

where L_s^η are the Lagrange interpolation coefficients determined from equations of type (2.27) but co-ordinate η is used instead of ξ and indexing changes: instead of i, index j; and $(f_p)i,j$ is the displacement (along axes x, y or z) of the j-th node falling on the i-th line along axis ξ [including the displacements of not only the selected points, but also those internal nodes are which displacements were determined using eqn. (2.28)].

The rotation angles of internal nodes are determined from eqns. of type (2.30) and (2.31) but co-ordinate η and index j are used instead of ξ and i respectively.

The method of interpolation by splines of the fourth and other orders is similar to the above procedure. Interpolation on lines parallel to axis ξ passing through point $\eta_j(j=0,\ldots,m)$ is carried out first.

A spline is constructed as follows on each line for $\xi\in\left[\xi_{i-d/2},\xi_{1+d/2}\right]$:

$$S^i\left(\xi,n_j\right)=\sum_{h=-d/2}^{d/2}L_{i-k}^\xi\left(f_p\right)_{i-k,j}, \qquad \ldots (2.36)$$

where d is the order of the polynomial;

$$L_{i-k}^\xi=\frac{\left(\xi-\xi_{i-d/2}\right)\left(\xi-\xi_{i-(d/2)}\right)\cdots\left(\xi-\xi_{i-(k-1)}\right)\cdots\left(\xi-\xi_{i-(k+1)}\right)\cdots\left(\xi-\xi_{i+d/2}\right)}{\left(\xi_{i-k}-\xi_{i-d/2}\right)\cdots\left(\xi_{i-k}-\xi_{i-(k-1)}\right)\left(\xi_{i-k}-\xi_{i-(k+1)}\right)\cdots\left(\xi_{i-k}-\xi_{i+d/2}\right)} \qquad \ldots (2.37)$$

$\left(f_p\right)_{i-k,j}$ are the displacements of the $(i-k)$-th estimated point on the j-th line in the direction of axis ξ (Fig. 2.8,b).

The following is the form of eqn. (2.29) for improving the accuracy of displacement values within the section between the i-th and $(i+1)$-th estimated points in a general case:

$$\left(f_p\right)_{s,j} = \frac{1}{d}\left[\left(f_p\right)_{s,j}^{i-\frac{d}{2}+1} + \cdots + \left(f_p\right)_{s,j}^{i+\frac{d}{2}}\right]. \qquad \ldots (2.38)$$

The rotation angles of internal nodes falling on lines passing through point $\eta_j(j=0,\ldots,m)$ are defined as:

$$\varphi'\left(\xi,\eta_j\right) = \sum_{k=-d/2}^{d/2} \theta_{i-k}^{\xi}\left(f_p\right)_{i-k,j} + \Delta\left(\xi,\eta_j\right), \qquad \ldots (2.39)$$

where $\qquad \theta_{i-k}^{\xi} = \dfrac{dL_{i-k}^{\xi}}{d\xi}$,

$\Delta(\xi,\eta_j)$ is an additional member allowing for variation in the co-ordinate systems xoy and $\xi 0'\eta$.

For plates, the additional member $\Delta(\xi,\eta_j)$ is determined once for axis ξ and axis η. For each node of a shallow shell, the additional member $\Delta(\xi,\eta_j)$ is determined separately since its value changes due to the non-zero curvature of the surface considered. In sections adjoining the contour, interpolation is done with Hermite splines using eqns. (2.32) to (2.34).

The following limitation should be borne in mind for the shell curvature comprising the cellular system: curvature of each shell along axes ξ and η is assumed such that interpolation along these axes can be done independently; i.e., such that the directrix of cosines of angles between the curvilinear axes ξ and η may be regarded as equal to one. Otherwise, interpolation should not be carried out and an equation for a two-dimensional interpolation constructed in a closed form, which is a very complex task.

To compare the accuracy of interpolation by the method developed, test calculations were made by the FEM for a square plate fixed along the contour by the proposed spline interpolation method and analytical method (Wilkinson, 1976; Marchuk, 1977). In the FEM method, the plate was divided into 100 square FEs. For calculations by the method developed, it was divided into four SEs, each into 25 square FEs. The disposition of points is shown in Fig. 2.11 and the types of loads on the plate in Fig. 2.12.

At the centre of the plate, no selected point was specially located so that deflection at the centre was determined by spline interpolation from the displacements of selected points. The deflection at the plate centre determined by the above three methods for each type of loading (Fig. 2.12) is given in Table 2.1. The results of test calculations show a fairly high accuracy of spline interpolation for the selected points.

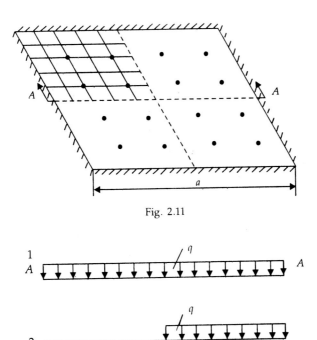

Fig. 2.11

1
A ↓↓↓↓↓↓↓↓↓↓↓↓↓↓↓↓↓↓↓ A q

2   ↓↓↓↓↓↓↓↓↓ q

3 ↑↑↑↑↑↑↑↑ ↓↓↓↓↓↓↓↓ q

Fig. 2.12

Table 2.1

No.	Solution by FEM	Solution by substructuring method with spline interpolation	Analytical solution
1	$1.288 \cdot 10^{-3}\lambda$	$1.277 \cdot 10^{-3}\lambda$	$1.28 \cdot 10^{-3}\lambda$
2	$6.44 \cdot 10^{-4}\lambda$	$6.39 \cdot 10^{-4}\lambda$	$6.40 \cdot 10^{-4}\lambda$
3	$3.07 \cdot 10^{-13}\lambda$	0	0

$\lambda = qa^4 / D$

2.4.2 Use of Piece-cubic Spline Functions

Lagrange interpolation polynomials discussed in Section 2.3.1 do not allow for the varying stiffness of sections and fix the rotation angles in the selected points. Cubic splines are free from this deficiency (Marchuk, 1977).

Let us examine a variant displacement field distribution based on piece-cubic splines assuming that the entire load has been replaced by a load statically equivalent to it and applied at the selected points.

In a general case, there are six degrees of freedom at each selected point of the fragment of cellular system: three displacements and three rotations. Then the displacement vector $\{q_p^I\}_{i,j}$ of the selected point falling on the intersection of the i-th and j-th lines of the dispersed network will assume the following form:

$$\{q_p^I\}_{i,j} = \left\{ (u_p^I)_{i,j}, (v_p^I)_{i,j}, (w_p^I)_{i,j}, (\varphi_p^I)_{i,j}, (\psi_p^I)_{i,j}, (\theta_p^I)_{i,j} \right\}, \qquad \dots (2.40)$$

where $(u_p^I)_{i,j}, (v_p^I)_{i,j}, (w_p^I)_{i,j}$ are the projections of linear displacements of the point on the common co-ordinate axes; $(\varphi_p^I)_{i,j}, (\psi_p^I)_{i,j}, (\theta_p^I)_{i,j}$ are the rotation at right angles to the surface of the I-th plate at the selected point i, j around the common co-ordinate axes x, y and z.

Let us determine the displacements along axis z of the internal nodes of the I-th plate on $(m + 1)$ lines parallel to axis ξ and passing through point $j(j = 0, m)$, having constructed for each of them a cubic spline $S_\xi^z(\xi, \eta_j)$ for which the conditions of agreement with the adjoining splines are fulfilled with respect to displacements and rotation angles at its ends:

$$S_\xi^z(\xi, \eta_j) = (w_p^I)_{i,j}; \qquad i = \overline{1, n}; \quad j = \overline{0, m}$$

for $\xi \in [\xi_{i-1}, \xi_i]$.

This spline may be represented as

$$S_\xi^z(\xi, \eta_j) = C_{i-1} \frac{(\xi_i - \xi)^3}{6h_i} + C_i \frac{(\xi - \xi_{i-1})^3}{6h_i} + \left[(w_p^I)_{i,j} - \frac{C_{i-1} h_i^2}{6} \right] \frac{(\xi_i - \xi)}{h_1} +$$

$$\left[(w_p^I)_{i,j} - \frac{C_i h_i^2}{6} \right] \frac{(\xi - \xi_{i-1})}{h_i}, \qquad \dots (2.41)$$

where $h_i = \xi_i - \xi_{i-1}$; $C_i = S''(\xi, \eta_j)$ from the spline construction.

Using the condition of the continuity of the first derivative of spline (3.41) at points $\xi_1, \dots, \xi_{n-1}$, we obtain $(n - 1)$ equations for determining the spline coefficients C_i. The remaining two equations can be obtained by using specific boundary conditions at the selected points of the spline: $i = 0$ and $i = n$. Thus, when the plate edges are hinged and points $i = 0$ and $i = n$ are located on them, the zero equality conditions of the second derivative of the spline at these points are used as the boundary conditions, i.e., $C_0 = 0$ and $C_n = 0$, since coefficients C_i of spline $S_\xi^z(\xi, \eta_j)$ represent in their form the second derivative of the spline in the dispersed point and are proportional to the deflection moments at these points.

In a general case, the remaining two equations are constructed based on the values of displacement and rotation angles at the boundary points are found in the initial computations of substructuring analysis, i.e., w_0, w_n, ψ_0 and ψ_n.

Having examined the single side (right) limit of the derivative $\left[S_\xi^z(\xi, \eta_j) \right]_\xi$ at the extreme point ξ_0, we find the following equation for the rotation angle at this point:

$$\left[S_\xi^z(\xi_0 + 0, \eta_j) \right]_\xi' = \psi_0 = -\frac{h_1}{3}C_0 - \frac{h_1}{6}C_1 + \frac{w_1 - w_0}{h_1}. \qquad \text{... (2.42)}$$

Thus, for splines on lines of mesh $\eta = \eta_j$, we have the following limiting equation:

$$2C_{0,j} + C_{1,j} + \frac{6}{h_1} \left[\frac{\left(w_p^I \right)_{1,j} - \left(w_p^I \right)_{0,j}}{h_1} - \psi_{0,j} \right]. \qquad \text{... (2.43)}$$

Similarly, on examining the single side (left) limit $\left[S_\xi^z(\xi_n + 0, \eta_j) \right]_\xi'$ at the edge point ξ_n, we derive the second limiting equation:

$$C_{n-1,j} - 2C_{1,j} = \frac{6}{h_n} \left[\psi_{n,j} - \frac{\left(w_p^I \right)_{n,j} - \left(w_p^I \right)_{n-1,j}}{h_n} \right]. \qquad \text{... (2.44)}$$

Eqns. (2.43) and (2.44) together with the equations emerging from the discontinuity conditions of the first derivative of spline (2.41) at points $\xi_{1,j}, \xi_{2,j}, \ldots \xi_{n-1,j}$, form system of equations relative to the spline coefficients C_{ij}:

$$
\begin{bmatrix}
2 & r_0 & & & & \\
d_1 & 2 & r_1 & & & \\
& \cdot & \cdot & \cdot & & \\
& & \cdot & \cdot & \cdot & \\
& & & \cdot & \cdot & \\
& & d_{n-1} & 2 & r_{n-1} \\
& & & d_n & 2
\end{bmatrix}
\begin{Bmatrix}
C_{0,j} \\ \cdot \\ \cdot \\ \cdot \\ \cdot \\ \cdot \\ \cdot \\ C_{n,j}
\end{Bmatrix}
=
\begin{Bmatrix}
P_0 \\ \cdot \\ \cdot \\ \cdot \\ \cdot \\ \cdot \\ \cdot \\ P_n
\end{Bmatrix}.
\qquad \text{... (2.45)}
$$

Here, $r_0 = 1$, $d_n = 1$, $r_k = \dfrac{h_k}{h_k + h_{k-1}}$, $d_k = 1 - r_k$, $k = \overline{1, n-1}$,

$$
P_0 = \frac{6}{h_1}\left[\frac{\left(w_p^I\right)_{1,j} - \left(w_p^I\right)_{0,j}}{h_1} - \phi_{0,j} \right],
$$

$$
P_n = \frac{6}{h_n}\left[\phi_{n,j} - \frac{\left(w_p^I\right)_{n,j} - \left(w_p^I\right)_{n-1,j}}{h_n} \right],
$$

$$
P_k = \frac{6}{h_k + h_{k+1}}\left[\frac{\left(w_p^I\right)_{k-1,j} - \left(w_p^I\right)_{0,j}}{h_{k-1}} - \frac{\left(w_p^I\right)_{k,j} - \left(w_p^I\right)_{k-1,j}}{h_k} \right]. \qquad \text{... (2.46)}
$$

The matrix of coefficients at C_{ij} is symmetrical with an exact diagonal predominance. According to the Hershgorin theorem on the localisation of typical values, it is non-specific and positive. Hence, there is only one solution to the system of eqn. (2.45) and for constructing the piece-cubic spline (2.41). The displacements of internal nodes on the j-th line parallel to axis ξ are found from this spline by substituting in it the actual values of co-ordinate ξ and displacements of selected (under consideration) points w_{ij} (Fig. 2.13).

The rotation angles at these nodes are determined by the equation for the first derivative of eqn. (2.41):

$$
\left[S_\xi^z(\xi, \eta_j) \right]_\xi^1 = C_{i-1,j}\frac{(\xi_i - \xi)^2}{2h_i} + C_{i,j}\frac{(\xi - \xi_{i-1})^2}{2h_i} + \frac{w_{i,j} - w_{i-1,j}}{h_i} - \frac{C_{i,j} - C_{i-1,j}}{6} + \Delta_{i,j}
$$

$$
\text{... (2.47)}
$$

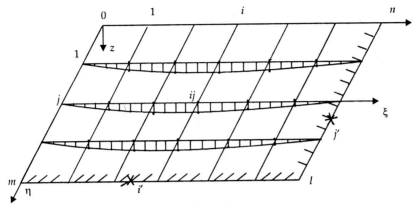

Fig. 2.13

for $\xi \in \left[\xi_{i-1}, \xi_1\right]$ where $\Delta_{i,j}$ is an additional factor which takes into account, as in eqn. (2.31), the variations between the general and local co-ordinate axes.

As in Section 2.3.1, assuming that each j-th line $\left(j=\overline{0, m}\right)$ in the direction ξ has on it $(l + 1)$ nodes of which $(n + 1)$ are selected (main) and the rest $(l - n)$ internal, displacements can be determined along axis z of internal (secondary) $(l - n)$ nodes on s lines of the mesh parallel to axis $\eta\left(i'=\overline{0,l}\right)$. For this purpose, let us construct the spline (Fig. 2.14) for

displacements $\left(w^l\right)_{s,j} = S_\eta^z(\xi_s, \eta_i)\left(i'=o,l; j=\overline{1,m}\right)$ found from eqn. (2.41) in

the first stage of spline interpolation (Fig. 2.13):

$$S_\xi^z(\xi_{i'}, \eta) = \tilde{C}_{j-1} \frac{(\eta_j - \eta)^3}{6h_j} + \tilde{C}_j \frac{(\eta - \eta_{j-1})^3}{6h_j} + \left[\left(w^1\right)_{i',j-1} - \frac{\tilde{C}_{j-1} h_j}{6}\right] \frac{(\eta_j - \eta)}{h_j} + $$

$$\left[\left(w^j\right)_{i',j} - \frac{\tilde{C}_j h_j}{6}\right] \frac{(\eta - \eta_{j-1})}{h_{j-1}}, \qquad \text{... (2.48)}$$

for $\eta \in \left[\eta_{j-1}, \eta_j\right]$, where $h_j = \eta_j - \eta_{j-1}$, $\tilde{C}_j = S_\eta''(\xi_{i'}, \eta)$ from the structure of the spline.

Coefficients $\tilde{C}_j$ of spline (2.48) are determined from the system of equations (2.45) after substituting the index n by m. Correspondingly, in eqn. (2.46), angle ψ_n should be substituted by φ_m:

$$\tilde{P}_{j=0} = \frac{6}{h_{j=1}} \left[\frac{\left(w^1\right)_{j=1,i'} - \left(w^1\right)_{j=0,i'}}{h_m} - \varphi_{j=0,i'}\right], \qquad \text{... (2.49)}$$

$$\tilde{P}_m = \frac{6}{h_m}\left[\varphi_{m,s} - \frac{\left(w^1\right)_{m,i'} - \left(w^1\right)_{m-1,i'}}{h_m}\right].$$

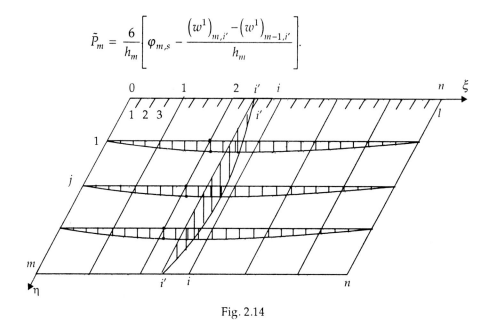

Fig. 2.14

The boundary conditions for determining coefficients $\tilde{C}_j$ of spline (2.47) are derived by analogy with eqns. (2.43) and (2.44):

$$2\tilde{C}_{j=1} + \tilde{C}_{j=1} = \tilde{P}_{j=0}, \tilde{C}_{m-1} + 2\tilde{C}_m = \tilde{P}_m. \qquad \qquad \dots (2.50)$$

After solving by trial and error the system of equations relative to coefficients $\tilde{C}_j$ using spline (2.47), displacements along axis z of all internal nodes on s lines $\left(s=0,\overline{1}\right)$ in the direction η are determined. Rotation angles $\varphi_{j,s}$ around axis x for internal nodes are determined using equations similar to eqn. (2.47) and derived after substituting in eqn. (2.47) the values of C and ξ by $\tilde{C}$ and η and indexes i and j by indexes j and i' respectively.

Spline interpolation of displacements along the other two axes is done similarly using eqns. (2.41) to (2.50) after appropriate substitution of notation for angular displacement.

Spline interpolation helps direct transition from the displacements of substructuring selected points of a high level to displacements of the original network of FE. Since the calculations are made using the same type of equations by resolving triple-diagonal systems of equations, a significant computer economy is resolved in the backward sweep of substructuring analysis.

It can be seen from the above discussion that linear displacements of all selected points and rotation angles at selected points at the plate boundaries are used for constructing a cubic spline. To construct Hermite interpolation polynomials, linear displacements as well as angular rotations at all selected points should bé known. Hence, when using Hermite polynomials, the error in determining the displacements of internal points will be somewhat less than when using interpolation by cubic splines.

The following are examples of analyses carried out to assess the accuracy of interpolation methodology of displacements by cubic splines in the backward sweep.

Example 2.1. An oblique-angled isotropic slab with inclination $\alpha = 45°$ fixed along the contour and subjected to a uniformly distributed load of intensity $\bar{q}$ (Fig. 2.15) has been studied. The Poisson ratio $v = 0$. For approximating the boundaries, let us use triangular FEs with nodes on the inclined edge which improves the accuracy of analysis (Maslennikov, *Substructuring Method in Structural Analysis*, 1979). The selected nodes are shown as dots in Fig. 2.15 (nodes 2, 5 and 8). The results are compared with the analyses carried out by A.M. Maslennikov (in: *Substructuring Method in Structural Analysis*, 1979) based on FEM and Vlasov's solution (1962-1964) based on variational methods.

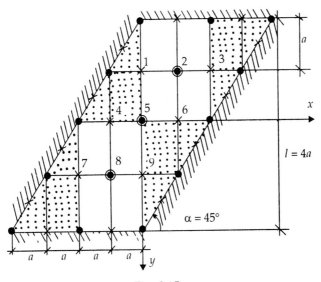

Fig. 2.15

Converting the distributed load into nodal loads, we obtain in the rectangular element node a load of $0.25\,qa^2$ and at a right angle to the triangular element $0.15\,qa^2$. As pointed out before (*Substructuring Method in Structural Analysis*, 1979) a plate with an orthogonal network of nodes was divided into 20 FEs and in substructure analysis into six SEs: two rectangular ones with four FEs each and four triangular with three FEs each.

Table 2.2 gives the deflection values and rotation angles at some nodes calculated by the FEM (*Substructuring Method in Structural Analysis*, 1979) and by interpolation. All deflections have a common multiplier qa^3/D and the rotation angles the common multiplier qa^4/D.

Table 2.2

No. of node	Degree of freedom	FEM*	Interpolation by cubic splines	Error Δ, %
1	2	3	4	5
1	ω	0.0749	0.0738	1.46
	φ_y	0.0876	0.0845	3.5
	φ_x	0.1287	0.1269	2.1
2	ω	0.0948	0.09476	0.04
	φ_y	−0.0365	−0.03622	0.76
	φ_x	0.0770	0.07653	0.61
3	ω	0.0327	0.0332	1.5
	φ_y	−0.656	−0.0632	3.6
	φ_x	−0.025	−0.0239	4.4
4	ω	0.0798	0.0786	1.5
	φ_y	0.1259	0.1223	2.8
	φ_x	0.0998	0.0966	3.2
5	ω	0.1533	0.15319	0.07
	φ_y	0.0	$0.7501 \cdot 10^{-7}$	–
	φ_x	0.0	$0.1778 \cdot 10^{-6}$	–

*Source: *Substructuring Method in Structural Analysis* (1979).

Figure 2.16 shows the deflection diagrams along lines 1–1 and 2–2 in the form of a continuous line as recorded by Vlasov (1962–1964) and as a broken line obtained by spline interpolation.

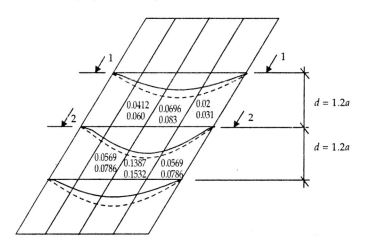

Fig. 2.16

Since in Vlasov's method, displacement values were determined for lines 1–1 not passing through the nodes of the FE network, suffice it to substitute pitch h in spline (2.41) by h/K, where K is the pitch reduction coefficient to obtain displacement values on this line by spline interpolation. The maximum deflection obtained based on Vlasov's variational method is $0.138713\ qa^4/D$, which differs from the calculated value by 9.4%.

The analysis shows that the deflections and rotation values calculated by spline interpolation are comparable in accuracy with the values obtained by the FEM (*Substructuring Method in Structural Analysis*, 1979).

Example 2.2. A cellular system, each plane of which is assumed as an independent SE, was analysed (Fig. 2.17). The division into FEs is shown on the frontal face of the cellular systems and nodes of the dispersed network of selected points marked.

The structural parameters selected are as follows:

$$\alpha = a/\delta = 80; \quad \beta = \delta^2 E/P = 5.25; \quad v = 0.3; \quad \text{and } a = 4 \text{ m}.$$

Two types of loading were analysed. In the first case a vertical load was applied along the two upper ribs (Fig. 2.17). Table 2.3 gives the displacement

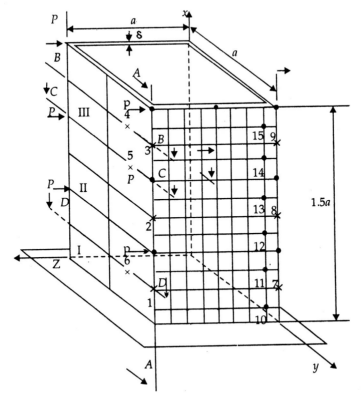

Fig. 2.17

values of internal nodes calculated by Hermite interpolation for lines A–A, B–B, C–C and D–D compared to the displacements in the original network of nodes. When using the FEM, the system of equations consisted of 1288 equations which dropped to 132 when using selected points.

Figure 2.18 shows the displacements of sections 1, 2 and 3 in plane *xoz* obtained by the spline interpolation method. In the second case the vertical load was applied on a single upper rib.

Figure 2.19 shows the displacements of upper and lower ribs and Fig. 2.20 the displacements of sections 1, 2 and 3 in the plane *xoz*.

Figure 2.21 shows the errors of vertical deflections obtained by the spline interpolation method for the second case of loading relative to the results of FEM analysis and Fig. 2.22 similar errors of calculating the rotation angles.

Table 2.3

Section	Node	Degree of freedom	FEM	Hermite spline	Error Δ, %
A–A	1	w	−0.0238	−0.0244	2.2
		φ_y	−0.0385	−0.0394	2.25
	2	w	−0.0738	−0.0838	8.9
		φ_y	−0.0245	−0.0249	1.2
	3	w	0.1187	0.1270	7.0
		φ_y	−0.0225	−0.0216	3.8
B–B	4	w	−0.1259	−0.1246	1.0
		φ_y	−0.0045	−0.0049	8.0
C–C	5	w	−0.1028	−0.1028	1.3
		φ_y	0.0041	0.0043	4.8
D–D	6	w	−0.0181	−0.0188	3.8
		φ_y	0.006	0.0062	2.5

2.4.3 Interpolation of a Displacement Field in Cellular Systems having Plates with Step-variable Stiffnesses

Stiffness matrix formation for selected points in a cellular structure having plates with step-variable stiffness is done by the method incorporated in Section 2.1. Each section of the plate with stiffness varying from the adjoining one is regarded as a distinct SE and a transformation matrix [S] formed for it. This only results in increasing the SEs of the first level.

When carrying out spline interpolation of displacements in the backward sweep in equations given in Section 2.3, modifications should be made to account for the stiffness variations in the plate sections. Cubic splines and Hermite polynomials are most convenient for displacement field interpolation of plates having step-variable stiffness.

Let us examine the *I*-th plate of the cellular system having step-variable

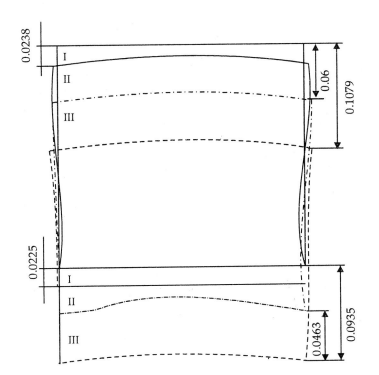

Fig. 2.18

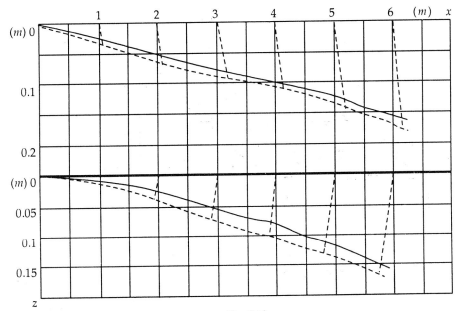

Fig. 2.19

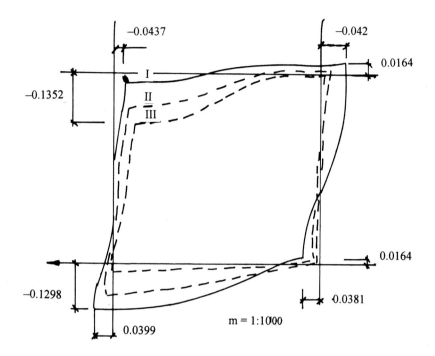

Fig. 2.20

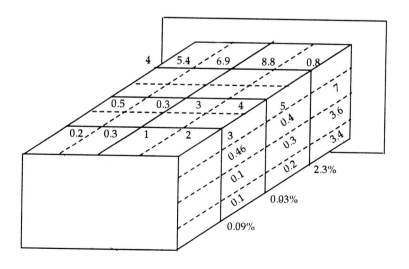

Fig. 2.21

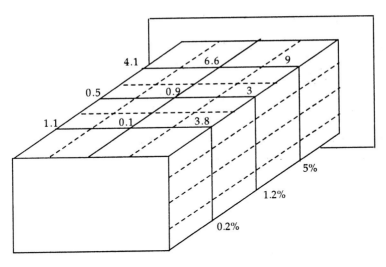

Fig. 2.22

stiffness along its length (width). As mentioned before, estimated points should necessarily be marked at the sites of stiffness variation.

Let the network of points have $(m + 1)$ lines along axis ξ and $(n+1)$ lines along axis η.

As in Section 2.3, let us use eqn. (2.41) to resolve the spline interpolation of displacements w in the lines parallel to axis ξ and passing through point η on the co-ordinate axis $\eta\left(j = \overline{0, m}\right)$. The limiting conditions of eqns. (2.43) and (2.44) allowing for stiffness variation of the sections are expressed in the following form:

$$2C_0 + C_1 D_1 D_0^{-1} = \frac{6}{h_1}\left[\frac{\left(w_p^I\right)_{1,j} - \left(w_p^I\right)_{0,j}}{h_1} - \psi_{0,j}\right],$$

$$C_{n-1} + 2C_n D_n D_{n-1}^{-1} = \frac{6}{h_n}\left[\psi_{n,j} - \frac{\left(w_p^I\right)_{n,j} - \left(w_p^I\right)_{n-1,j}}{h_n}\right], \quad \text{... (2.51)}$$

where $h_k = \xi_k - \xi_{k-1}$.

Here, the factors D_0, D_1, D_{n-1} and D_n represent cylindrical stiffness of corresponding sections of the I-th plate:

$$D_i = \frac{E_i \delta_i^3}{12\left(1 - v_i^2\right)}, \quad \text{.... (2.52)}$$

in which δ_i, E_i and v_i are the thickness, modulus of elasticity and the Poisson ratio of the i-th section of the plate.

The coefficients of the system of equations (2.45) relative to the unknown coefficients c_i and the values of the right part allowing for eqn. (2.51) differ from eqn. (2.46) and are as follows:

$$r_i = h_i \, D_{i+1} \left[D_i \left(h_i + h_{i-1} \, D_i \, D_{i-1}^{-1} \right) \right]^1 ,$$

$$d_i = h_{i-1} \left(h_i + h_{i-1} \, D_1 \, D_{i-1}^{-1} \right)^{-1} ,$$

$$P_i = \frac{6 \left[\left(w_p^I \right)_{i+1} - \left(w_p^I \right)_{i,j} \right] h_{i+1}^{-1} - \left[\left(w_p^I \right)_{i,j} - \left(w_p^I \right)_{i-1,j} \right] h_i^{-1}}{h_i + h_{i-1} \, D_i \, D_{i-1}^{-1}} , \qquad \dots \ (2.53)$$

$$P_n = \frac{6}{h_n} \left[\psi_{n,j} - \frac{\left(w_p^I \right)_{n,j} - \left(w_p^I \right)_{n-1,j}}{h_n} \right] \left(\frac{D_{n-1}}{D_n} \right) .$$

Taking into consideration the varying stiffness, eqns. (2.49) and (2.50) assume the following form similar to eqn. (2.51):

$$2\tilde{C}_{j=0} + \tilde{C}_{j=1} \, D_{j=1} \, D_{j=0}^{-1} = \frac{6}{h_{j=1}} \left[\frac{\left(w^1 \right)_{j=1,i} - \left(w^1 \right)_{j=0,i'}}{h_j} - \varphi_{j=0,i'} \right] \qquad \dots \ (2.54)$$

$$\tilde{C}_{m-1} + 2\tilde{C}_m \, D_m \, D_{m-1}^{-1} = \frac{6}{h_m} \left[\varphi_{m,i'} - \frac{\left(w_p^I \right)_{m,i'} - \left(w_p^I \right)_{m-1,i'}}{h_m} \right] .$$

Example 2.3. To evaluate the accuracy of the cubic spline interpolation method when analysing thin-walled cellular systems with plates having step-variable stiffness, a test analysis was carried out on a rectangular plate fixed along the contour. The plate sections vary in stiffness as follows: the left half of the plate has a modulus of elasticity twice that of the right half (Fig. 2.23).

In the FEM analysis, the plate was divided into 100 square FEs and in substructuring analysis into four SEs. The arrangement of the selected points is shown in Fig. 2.23 and the types of loading on the plate in Fig. 2.24. Deflection diagrams obtained by spline interpolation based on displacements at selected points are shown in Fig. 2.25.

The deflection at the centre of plate determined by the FEM and using spline interpolation is given for the two types of loading in Table 2.4. The results show that the spline interpolation based on the selected points even in the case of step-variable stiffness of plates ensures quite a high analytical accuracy.

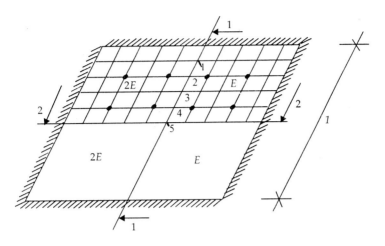

Fig. 2.23

Table 2.4

Loading scheme	Node	Deflection ratio $(w = w/\lambda)$		Error Δ, %
		FEM	Spline interpolation	
	1	0.0634	0.0656	3.5
	2	0.1955	0.1967	0.6
1	3	$1.808 \cdot 10^{-7}$	$1.184 \cdot 10^{-7}$	—
	4	0.2053	0.2079	1.2
	5	0.0796	0.0811	1.9
	1	0.0764	0.0787	3.0
	2	0.3941	0.3961	0.5
2	3	0.4163	0.4147	0.4
	4	0.1962	0.198	0.9
	5	0.0331	0.0342	0.3

$$\lambda = \left(ql^4/D\right) 10^{-3}$$

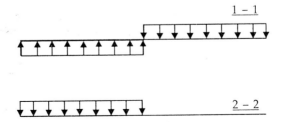

Fig. 2.24

Example 2.4. As an example of spline interpolation, the cellular system depicted in Figure 2.26 was analysed. On the front and upper faces, nodes on the network of dispersed selected points are marked as points. The global system of co-ordinates x, y and z is shown in the Figure. The system parameters are: plate thickness $\delta = 0.05$ m and the Poisson ratio $v = 0.3$. The number of degrees of freedom in each node of the cell is six: u, v, w, φ_y, φ_x and φ_z. Plate sections varying in stiffness are numbered.

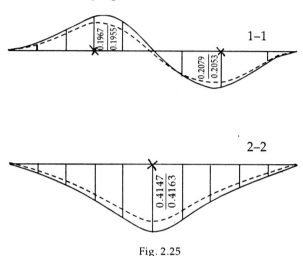

Fig. 2.25

The following values for the modulus of elasticity were selected for analysis :

1) 2.10×10^5 MPa;
2) 1.68×10^5 MPa;
3) 1.54×10^5 MPa;
4) 1.38×10^5 MPa;
5) 1.10×10^5 MPa;
6) 1.03×10^5 MPa.

Table 2.5 shows a comparison of the results of analysis with FEM analysis for the given sections.

The interpolation error increases close to the nodes. Table 2.6 shows the displacements of crossed nodes on line C–C determined by the cubic splines method. Rotations of the extreme selected points on line C–C around axis y in the initial computation were taken as angles $\varphi_{o,j}$ and $\varphi_{n,j}$ in the limiting conditions (2.51).

Table 2.5

Section	Degree of freedom	Node	FEM	Cubic splines	Error Δ,%	Hermite polynomial	Error Δ,%
A–A	w	1	−0.1105	−0.1122	1.5	−0.1123	1.6
	φ_y		0.0082	0.0083	1.17	0.00825	0.6
	w	2	−0.1076	−0.1096	1.8	−0.10965	1.8
	φ_y		−0.0036	−0.0040	11	−0.0039	8.3
	w	3	−0.0151	−0.0149	1.1	−0.01485	1.6
	φ_y		−0.0266	−0.0265	0.36	−0.0264	0.6
B–B	w	4	−0.0786	−0.0777	1.1	−0.0781	0.7
	φ_y		−0.0304	−0.0302	0.7	−0.0304	0.0
	w	5	−0.1366	−0.1369	0.2	−0.1366	0.0
	φ_y		−0.0303	−0.03012	0.58	−0.0304	0.3

Table 2.6

Degree of freedom	Node	FEM	Cubic splines	Error Δ, %
w	6	−0.0184	−0.0191	3.8
φ_y		−0.0059	−0.00617	4.5
w	7	−0.0182	−0.0189	3.8
φ_y		−0.0059	−0.0061	3.7

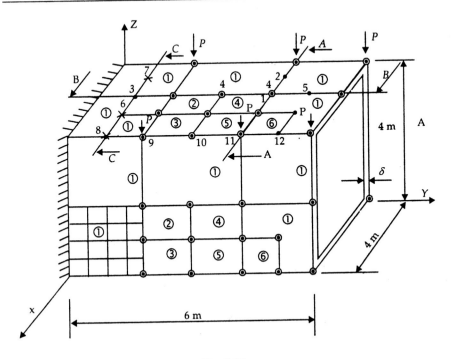

Fig. 2.26

Displacements w of the upper rib under load are shown in Table 2.7. Interpolation was by cubic splines and Hermite polynomials. The results are compared with the FEM solution in a dense network of nodes. The Table shows that the interpolation error rises on the loaded lines and close to the zones falling under the influence of nodes.

Table 2.7

Node	FEM	Cubic interpolation	Error Δ, %	Hermite polynomial	Error Δ, %
8	−0.01097	−0.00893	18.5	−0.01219	11.1
9	−0.04158	−0.042557	3.36	−0.04347	4.5
10	−0.06483	−0.067248	3.7	−0.06682	3.0
11	−0.1047	−0.10157	2.9	−0.10169	2.8
12	−0.1245	−0.1229	1.3	−0.12371	0.63
13	−0.1494	−0.13994	6.3	−0.15176	1.5

The above examples demonstrate the fairly high degree of accuracy in determining the internal nodal displacements by cubic spline interpolation method when there is step-variable stiffness in the plate system.

2.5 Enlarging of Number of Selected Points in Areas of High Stress Gradient

The interpolation procedure of a displacement field described in Section 2.4 for all the plates of a cellular system does not permit the precise description of zones with a high anticipated stress gradient: near openings, local clamps, sites of intense load variation and the like.

A rational enlarging scheme for the network of selected points is examined in this section. When analysing a cellular system in general, a extended network with fairly large cells is formed. After determining the displacement at selected points in the entire cellular system and following the spline interpolation procedure by the method described in the preceding section, the most loaded plate sections are identified. A new network of selected points is marked in these sections in a sequential decreasing of superelements size order (Fig. 2.27).

The first two or three rows of selected points on the edges of openings or at the position of an applied clamp fall on all the nodes of the basic FEs. These are followed by rows of larger SEs (with length of side doubled as shown in Fig. 2.27, or more). These SEs are initially analysed individually and their stiffness matrix formed. Along with estimated points, boundary points are also identified to ensure coupling of the various sizes of SEs (Fig. 2.28).

Initially, the internal nodes within each SE are excluded. Later, the SEs are formed around the dense network of selected points in the form of boundary rows. The boundary nodes in the coupling zone with the SE of the preceding row are progressively eliminated. This procedure helps retain a system of equations of relatively low order by closely dividing the zone into basic FEs. The primary conditions are taken into consideration when resolving the system of equations displacements and rotations of external boundaries of zones (marked as double lines in Fig. 2.27). These displacements and rotations are defined by spline interpolation in the stage of general analysis of the structure.

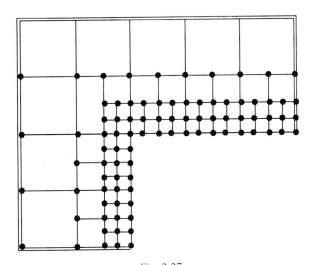

Fig. 2.27

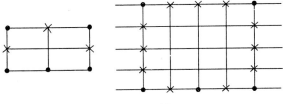

Fig. 2.28

Since all the nodes of basic FEs remain intact on the edge of a stress concentration (opening, clamping) in the maximum stress zone, a backward sweep of the substructuring procedure is not necessary for defining the maximum stresses: the displacement of these FEs are known from the resolution of the system of equations. When it is required to determine the complete displacement field in the initial FE net, the spline interpolation procedure is followed by the methodology described in the preceding section. It must only be remembered that the number of estimated points on each line and the line length are different.

It is not required that areas of stress concentrations be rectangular. Non-rectangular areas are readily modelled for any shape by introducing triangular basic FEs. Figure 2.29 shows an example of such division.

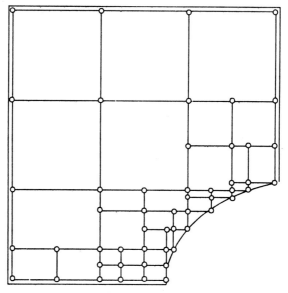

Fig. 2.29

2.6 Determination of Stresses and Comparison with Experimental Results

Stresses are determined after defining all the internal displacements in the l-th plate of the cellular system.

Let the l-th plate under consideration fall in the plane xoy and the network of all the nodes in it have $(l+1)$ lines on x axis and $(k+1)$ lines on y axis. Let us isolate a distinct cell of the nodal network formed by the intersection of the j-th and $(j+1)$-th lines on x axis with the i-th and $(i+1)$-th lines on y axis and regard it either as a single tetragonal FE or as two triangular FEs (Fig. 2.30).

By knowing the pitch along axes x and y, stiffness parameters and the thickness of the l-th plate, let us formulate the matrix of the FE forces which in most cases coincides with the FE stiffness matrix. The stiffness matrix of a tetragonal or triangular planar FE of a cellular structure can be formed for example by combining the stiffness matrix of the FE functioning in its plane with that of a bending finite element as carried out by Postnov (1977).

Later, from among the FE nodes (Fig. 2.30), the displacement vectors corresponding to nodes q_{ij}, $q_{i+1, j}$ and $q_{i+1, j+1}$ are selected from the displacement matrixes of the l-th plate.

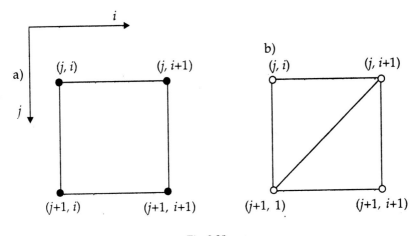

Fig. 2.30

The nodal forces in the FE are determined by multiplying the stifness matrix of FE by the displacement vector of its nodes:

$$\{S_R\} = [K_R]\{q_{i,j} \,; q_{i+1,j} \,; q_{i,j=1} \,; q_{i+1,j+1}\},$$ ... (2.55)

where S_R is the vector of FE nodal forces.

After defining the forces in all the FEs adjoining the $k(i,j)$-th node (Fig. 2.31), stresses (forces) in it are determined using the well-known equations of Sapozhnikov and Gorelov (1982):

$$\sigma_x = \frac{S_{14,I} + S_{19,II}}{b \cdot \delta}, \qquad \sigma_y = \frac{S_{20,I} + S_{10,III}}{a \cdot \delta},$$

$$\tau_{xy} = \frac{S_{14,II} + S_{4,II}}{a \cdot \delta}, \qquad \tau_{yx} = \frac{S_{10,III} + S_{5,IV}}{b \cdot \delta},$$

$$M_x = \frac{1}{b}\left(S_{17,I} + S_{12,II}\right), \quad M_y = \frac{1}{b}\left(S_{18,I} + S_{8,III}\right), \qquad \text{... (2.56)}$$

$$M_{x,y} = \frac{1}{b}\left(S_{12,II} + S_{2,IV}\right) \quad M_{y,x} = \frac{1}{b}\left(S_{18,I} + S_{13,II}\right),$$

$$Q_x = \frac{1}{b}\left(S_{6,III} + S_{1,III}\right) \quad Q_y = \frac{1}{b}\left(S_{16,I} + S_{6,III}\right),$$

where a and b are the FE dimensions and δ its thickness.

To evaluate the accuracy of stress determination by this method, the cellular structure covered in the second example (Section 2.4.2) was analysed for the first type of loading.

The stress values calculated for points 10 to 15 (Fig. 2.17) and their percentage deviation from the FEM analysis are shown in Table 2.8.

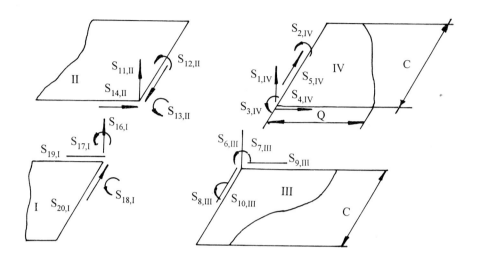

Fig. 2.31

Table 2.8

| Node | σ_x | | Error | $\tau_{x,y}$ | | Error |
	FEM	Splines	Δ, %	FEM	Splines	Δ, %
10	−186.264	−198.38	6.5	−141.497	−150.67	6.4
11	−116.379	−122.52	5.2	−117.47	−121.31	3.3
12	−68.091	−71.72	5.3	−103.042	−106.53	3.4
13	−34.349	−35.93	4.6	−84.854	−88.92	4.7
14	−12.276	−12.98	4.9	−58.66	−60.07	2.4
15	−1.901	−2.01	5.7	−21.186	−22.17	4.6

In FEM analysis, the moments in a given node are determined individually for each FE and these values generally differ. The average value of nodal moments of FEs adjoining the nodes are adopted as the final moment value at the node. If the FEs adjoining the node have varying stiffness values, the nodal moment is determined as the sum of nodal moments multiplied by the distribution coefficients (*Postnov, Method in Structural Analysis*, 1979):

$$\alpha_i = \frac{r_i}{\sum_{j=1}^{n} r_i}, \qquad \ldots (2.57)$$

where n is the number of elements adjoining the node and r_j is the reaction of the j-th element in the link at this node during one rotation.

In a general case, coefficients α_i of moments relative to different axes may differ.

To avoid the cumbersome FEM calculation of forces and also to eliminate incongruencies in the moment diagram and average values of nodal moments calculated by the aforesaid method, the coefficients of the constructed cubic spline (2.41) can be used. Structurally it represents the second derivative in the nodes of a enlarged network of selected points, i.e., values defining the deflection moments:

$$M_x = D_\xi \frac{\partial^2 S}{\partial \xi^2} + D_\mu \frac{\partial^2 S}{\partial \eta^2}, \qquad \ldots (2.58)$$

$$M_y = D_\mu \frac{\partial^2 S}{\partial \eta^2} + D_\eta \frac{\partial^2 S}{\partial \eta^2}. \qquad \ldots (2.59)$$

For angle i on the j-th line parallel to the axis, spline coefficients are determined by the equation:

$$\frac{\partial^2 S}{\partial \xi^2} = \left[s^z \left(\xi, n_j \right) \right]''_{\xi\xi} = \left[C_{i,j} \right] \text{ at } \xi = \xi_i. \qquad \ldots (2.60)$$

A similar equation is available for spline coefficients on the i-th line parallel to axis η:

$$\frac{\partial^2 S}{\partial \eta^2} = \left[s^z \left(\xi_i, \eta \right) \right]''_{\eta\eta} = \left[C_i \right] \text{ at } \xi = \xi_i. \qquad \ldots (2.61)$$

By substituting in the above equation $\eta = \eta_j$, the values of $\tilde{C}_{i,j}$ can be found for node j on the i-th line parallel to axis η.

Thus, the approximate value of moments M_x and M_y in the selected nodes can be determined from the values of spline coefficients without resorting to using eqns. (2.56).

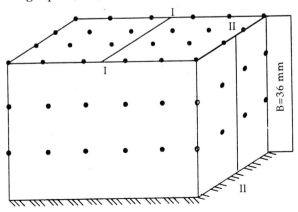

Fig. 2.32

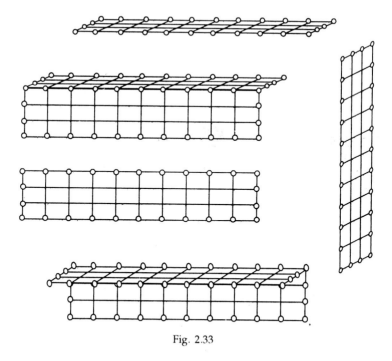

Fig. 2.33

For experimental assessment of stress determination by the above-developed method using eqns. (2.56), a thin-walled cellular shell was analysed. Novitskii (Novitskii, 1984) tested a model of it for internal pressure.

In Novitskii's experiment, the shell model was a thin-walled rectangular shell without a bottom, size 500 × 360 × 360 mm and wall thickness 12 mm. The stresses arising in the shell walls were measured by strain gauges of base 5 mm and resistance 100 ohms. A 45° rosette was used to determine the main stresses in the wall.

The free edges of the model were firmly fixed by welding to a thick steel plate.

The cellular shell was tested for an internal pressure of 1 MPa. The location of the selected points is shown in Figure 2.32.

The basic FEs selected were 40 × 50 mm and 40 × 40 mm in size. The stiffness matrices of selected points were formulated at three levels. Figure 2.33 shows the SEs of the first level, reduction of stiffness matrix and their loading using the transformation matrix [S].

Figure 2.34 shows SEs of the second level which, in the third level, were joined to form a full structure.

After solving the system of equations and determining the displacements at selected points, the internal nodal displacements were interpolated using splines based on Lagrange and Hermite polynomials. Later, the stresses in

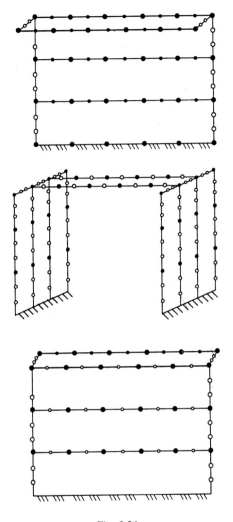

Fig. 2.34

the FEs were determined. The main stress diagrams for sections 1–1 and 2–2 (Fig. 2.32) are shown in Figures 2.35 and 2.36. The circles in the diagrams depict Novitskii's experimental results.

A comparison of the experimental stress values with the results of analysis by the method developed reveals their good accord. The maximum deviation of stress values obtained analytically from the experimental results was 8.57%.

Such deviations are noticed in the zone of low stresses (σ in Fig. 2.36) of four ribs; the corresponding maximum deviation in the zone of significant

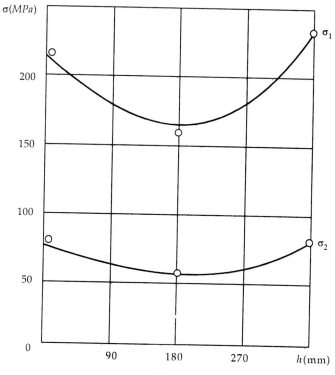

Fig. 2.35

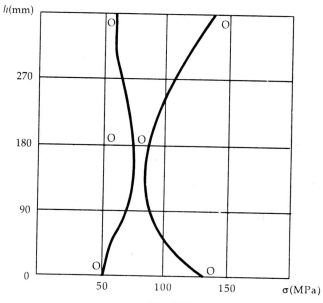

Fig. 2.36

stresses was 4.8%. Considering that the scatter of experimental values per se goes up in this zone to 3.6%, the result should be regarded as satisfactory. It may be pointed out that minimum stresses arise in the shell corners and maximum in section 1–1 on the long ribs of the shell, this too agrees with the experimental results.

Examples of using the developed substructuring methodology for analysing complex cellular systems using spline interpolation in the backward sweep are cited (Ignat'eva 1983; Ignatiev and Gorelov, 1985, 1986; Ignatiev and Kaurova, 1989).

References

*Aleksandrov, A.V., B.Ya. Lashchenikov and N.N. Shaposhnikov. 1983. *Structural Engineering. Thin-walled Spatial Systems*. Stroiizat, Moscow, 488 pp.

*Bate, K. and E. Wilson. 1982. *Numerical Methods of Analysis and Finite Element Method*. Stroiizdat, Moscow, 466 pp.

Bathe, K.J., and E.L. Wilson 1976. *Numerical Analysis of Finite Element Analysis*. Englewood Cliffs NJ: Prentice-Hall.

*Bezukhov, N.I., O.V. Luzhin and N.V. Kolkunov. 1987. *Structural Stability and Dynamics*. Vyssh. Shk., Moscow, 264 pp.

*Berezin, I.S. and N.P. Zhidkov. 1966. *Computational Methods*. Nauka, Moscow. 1: 632.

Birkhoff, G. and H.L. Garabedian. 1960. Smooth surface interpolation. *J. Math. Phys.*, 39: 353–368.

*Blokhina, I.V. 1989. Development of Substructuring Method of Static and Dynamic Analysis of Thin-walled Cellular Systems. Cand. Diss. Technical Sciences. Volgogradsk, Inzh.–Stroit. Int, Volgograd, 28 pp.

*Blokhina, I.V. and O.M. Ignat'eva. 1987. Substructuring Splines for Analysing Planar Systems for Dynamic Influences. Volgogradsk Inzh.-Stroit. In-t, Volgograd, 9 pp. Deposited in VINITI, 22 October, 1987, No. 7437.

Craig, R.R. and M.C.C. Bampton. 1968. Coupling of structures for dynamic analysis. *AIAA J.*, 6.

*Deak, A.L. and T.H.H. Pian. 1967. Supplementing the smooth interpolation methods of surfaces to analysing structures based on discrete scheme. *Raketnaya Tekhnika i Kosmonavtika*: 5(1): 235–237.

*Dynamic Analysis of Buildings and Structures. Designers' Handbook. 1984. Stroiizdat, Moscow, 303 pp.

*Finite Element Method in the Mechanics of Solid Bodies. 1982. A.S. Sakharov and I. Al'tenbakh (eds.) Vischcha Shk., Kiev, 480 pp.

*Grinenko, N.I. and V.V. Mokeev. 1985. Problems of studying structural vibrations by finite element method. *Prikladnaya Mekhanika*, 21(3): 12–15.

*Ignatiev, V.A. 1973. *Analysis of Periodic Pivotal Systems*. Saratov, 427 pp.

*Ignatiev, V.A. 1981. *Superquantification Methods for Analysing Complex Pivotal Systems*. Izd-vo SGU, Saratov, 108 pp.

*Ignatiev, V.A. and S.F. Gorelov. 1985. Analysis of Cellular Systems Using Spline-substructuring. Volgogradsk. Inzh.-Stroit. In-t, Volgograd, 25 pp. Deposited in VINITI, 15 March, 1985, No. 2183.

*Ignatiev, V.A. and S.F. Gorelov. 1986. Analysis of cellular systems by substructuring method with spline-interpolation of transfers. *Izv. Vuzov. Ser. Stroitel'stvo i Arkhitektura*, 12: 18–24.

*Ignatiev, V.A. and O.M. Ignat'eva. 1987. Interpolational Determination of Transfer Fields and Stresses in Substructuring Analysis of Thin-walled Cellular Systems.

56

Volgogradsk. Inzh-Stroit. In-t, Volgograd, 41 pp. Deposited in VINITI, 24 June 1987, No. 4595.

*Ignatiev, V.A. and T.M. Kaurova. 1989. Substructuring Static Analysis of Cellular Systems Using Spline-interpolation in Forward and Backward Sweeps. Volgogradsk. Inzh.-Stroit. In-t, Volgograd, 68 pp. Deposited in VINITI. 11 July 1989, No. 4337-B89.

*Ignat'eva, O.M. 1983. Use of Lagrange interpolation polynomials for analysis by dispersed network method. *Prikladnaya Teoriya Uprugosti*, SPI, Saratov, pp. 72–26.

*Ivanteev, V.I. and V.D. Chuban'. 1984. Form and frequency analysis of free vibrations of structures by multilevel dynamic condensation method. *Uchenye Zapiski TsAGI*, 15(4): 81–92.

*Kiselev, V.A. 1973. *Plate Analysis*. Stroiizdat, Moscow, 151 pp.

*Krjukov, N.N. Analysis of oblique-angled and trapeziform plates by spline-function method./Appl. Mech. Kijev-1997-33. N5-pp77-81 (Russ).

*Marchuk, G.I. 1977. *Methods of Computational Mathematics*. Nauka, Moscow, 454 pp.

*Maslennikov, A.M. 1987. *Structural Analysis by Numerical Methods*. Izd-vo LGU, Leningrad, 224 pp.

*Norri, D. and de Friz, Zh. 1981. *Introduction to Finite Element Method*. Mir, Moscow, 300 pp.

*Novitskii, S.V. 1984. Identifying Hazadrous Zones in Cellular-type Shells. MIIT, Moscow, 8 pp. Deposited in VINITI, 20 May 1984, No. 8270-84.

*Petrov, V.V. 1975. *Method of Successive Loadings in Non-linear Theory of Plates and Shells*. Izd-vo SGU, Saratov, 119 pp.

*Petrov, V.V., I.P. Ovchinnikov and V.I. Yaroslavskii. 1976. *Analysis of Plates and Shells Made of Non-linear Elastic Materials*. Izd-vo SGU, Saratov, 133 pp.

*Pletnev, V.I. 1983. Analysis of cellular systems by the method of forces and method of transfers together with finite element method. In: *Structural Engineering*. Izd-vo LISI, pp. 26–32.

*Postnov, V.A. 1977. *Numerical Methods of Analysing Ship Structures*. Sudostroenie, Leningrad, 280 pp.

*Postnov, V.A. and I.Ya. Kharkhurim. 1974. *Finite Element Method for Analysing Ship Structures*. Sudostroenie, Leningrad, 344 pp.

*Rice, J. 1984. *Matrix Calculations and Mathematical Procedures*. Mir, Moscow, 264 pp.

Rohr, U. 1985. Lokale finite Elementnetzverfeinerung bei Platten- und Scheibenaufgaben mittels gemischter Interpolation. *Schiffbauforschung*, 24(1): 39–50.

*Sapozhnikov, A.I. 1980. Contour points method for analysing structures. In: *Structural Engineering and Analysis*, 5: 59–61.

*Sapozhnikov, A.I. and S.F. Gorelov. 1982. Structural analysis by finite element method with phased formation of stiffness matrix. In: *Structural Engineering and Analysis*, 4: 54–56.

*Segerlind, L. 1979. *Application of the Finite Element Method*. Mir, Moscow, 392 pp.

*Shoop, T. 1982. *Solving Engineering Problems in a Computer*. Mir, Moscow, 238 pp.

*Smirnov, A.F., A.V. Aleksandrov, B.Ya. Lashchenikov and N.N. Shaposhnikov. 1984. *Structural Engineering. Dynamics and Stability of Structures*. Stroiizdat, Moscow, 415 pp.

Substructuring Method in Structural Analysis. 1979. V.A. Postnov (ed.), Sudostroenie, Leningrad, 288 pp.

*Timoshenko, S.P. and S. Voinovskii-Kriger. 1963. *Plates and Shells*. Fizmatgiz, Moscow, 636 pp.

Tongchen Miao, Wang Wei, Qu Jingdong. The spline finite-point method for dynamic analysis of arbitrary tetragon plate. Zhendong yu chongji=Vibr. and Shock.-1996-15.N4-p.p. 59-62. (China).

*Vlasov, V.Z. 1962–1964. *Selected Works*. 3 vols. Izd-vo AN SSSR, Moscow, vol. 1, 528 pp; vol. 2, 507 pp; vol. 3, 472 pp.

*Voronenok, E.Ya. and S.V. Sochinskii. 1981. Interpolation condensation of stiffness matrix when solving structural engineering problems by substructuring method. Prikladnaya Mekhanika, 17(6): 114–118.

*Wilkinson, R. 1976. *Handbook of Algorithms in ALGOL. Linear Algebra*. Mashinostroenie, Moscow, 390 pp.

*Zav'yalov, Yu.S., B.I. Kvasov and V.L. Miroshnichenko. 1980. *Spline Function Methods*. Nauka, Moscow, 352 pp.

*Zenkevich, O. and K. Morgan. 1986. *Finite Elements and Approximation*. Mir, Moscow, 318 pp.

Zivorad Bojovic, Grbič Dragolub. Splajn-funkcija petog zeda./Tehnika-1997-52, N3-4.- C.NG1-NG3.(serb.-chozv.)

*Asterisked entries are in Russian—General Editor.

3

Statics of Prismatic and Cylindrical Shells of Multiconnected Section with Periodic Structure

3.1 Introduction

V.Z. Vlasov (1958) introduced a general theory in the form of the variational method of displacements for analyzing prismatic shells, taking into account the deformability of the section contour. In shells of medium length, only the longitudinal and shear forces and transverse bending moments together with their intersecting forces are taken into consideration from among the internal forces while the impact of longitudinal bending moments and torsional moments is regarded as negligibly small. Displacements of any point on the middle surface of the shell are represented in the form of finite expansion vectors, as a result of which the problem is reduced to a system of ordinary differential equations of order $(m + n)$ in which m and n are the number of the degrees of freedom of an elementary strip or frame isolated from the shell by two congruent cross-sections, namely from the plane and in the plane of the cross-section. The order of the system of equations of Vlasov's method generally exceeds by a few times the number of finite elements in a multiconnected section and can therefore attain a significant value. The difficulties associated with these calculations increase many times when studying the behaviour of structures under the influence, e.g., of a moving concentrated load. Such problems for shells of multiconnected sections have thus remained unsolved until recently (*Structural Mechanics in the USSR*, 1969). The order of the system of equations in Vlasov's method can be reduced firstly by reducing the number of given functions of transverse and longitudinal displacements $\varphi_i(s)$ and $\psi_k(s)$. For the class of shells under consideration, a multiconnected section with a periodic structure, such a possibility is available relative to the family of functions $\psi_k(s)$. The fact of periodicity helps take into account only the contour deformation instead of a finite multiplicity of approximating functions describing the displacements in the plane of the cross-section of the multiconnected shell. Deriving it can be comparatively simple by the static method from a solution to the problem of cylindrical bending or bending of an elementary strip of frame isolated from an actual shell under the action of load comparable to the given loading. When resolving this subsidiary problem, the use of a mathematical system of equations in finite differences, which is extremely effective in the case of periodic problems, enables finding the unknown function in a closed analytical form. A natural consequence of this approach is a reduction in the order of the system of equations in Vlasov's method to $(m + 1)$.

The next significant simplification involved ignoring shear strains when analysing prismatic shells for the action of flexural loads. Mileikovskii (1950, b) derived a system of differential equations of the variational method for shells of a multiconnected section, ignoring shear strains which lower the order of the system of equations to number m.

When analysing shells of a multiconnected section possessing a periodic structure, maximum simplification of the mathematical aspect of the problem can be achieved by ignoring the impact of shear strain and simultaneously considering the fact of periodic structure. By introducing a single function $\psi(s)$ instead of a multiplicity of functions $\psi_k(s)$, maximum simplification is possible in that the problem of flexure of a shell of a multiconnected section will invariably be described by a single differential equation of the fourth order irrespective of the number of cells in the shell section.

The approach described here helps derive in a closed analytical form a solution to a whole range of complex problems in structural mechanics involving shells of a multiconnected section which is very difficult within the framework of conventional methods due to difficulties such as constructing the surfaces falling under the influence of displacement forces and strains in spans of cellular bridges without diaphragms and with a multiconnected section (continuous or discontinuous), analysis of pontoons with a multiconnected section as shells in the elastic foundation to the action of variable loading, and the analysis of a ship's body with a multiconnected section with longitudinal orientation of compartments (Fig. 3.0). Figure 3.0 illustrates several examples of multicell plate structures. These problems are solved with due consideration of the contour deformability of the shell section.

The developed method can effectively solve the problems of analysing cylindrical double-walled shells of tanks under the action of internal hydrostatic pressure as well as lateral concentrated loading.

This opens up the possibility of solving a large series of flexural problems of rectangular flat and cylindrical three-layered panels with a ribbed filler, also taking into account the discrete nature of the filler. Such solutions are needed when the prevailing approximate theories of three-layered structures produce unreliable results.

The advantages of the proposed method of analysis lies in that it eliminates the difficulties associated with formulating the boundary conditions on the linear edges of the shell as these conditions are automatically satisfied in the static method of approximation.

60

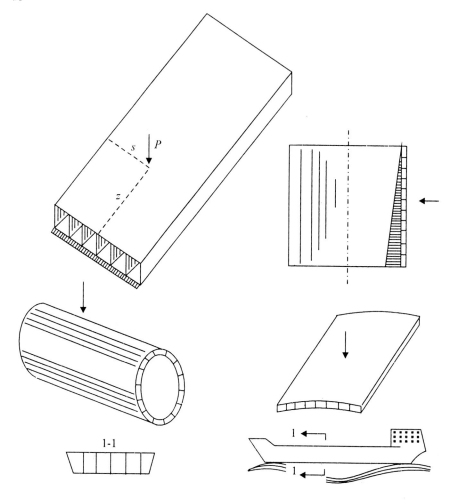

Fig. 3.0

3.2 Analysis of Prismatic Shells of Multiconnected Section with Periodic Structure Allowing for Shearing Strains

3.2.1 *Variational Method of V.Z. Vlasov for Analysing Prismatic Shells of Medium Length*

There are two main directions in a shell of multiconnected section: lateral (s) and longitudinal (z). Any point on the midsurface of any member of the shell may undergo longitudinal displacement $u = u(z, s)$ and lateral tangential displacement $v = v(z, s)$. From among the inner forces, those retained are as shown in Figure 3.1.

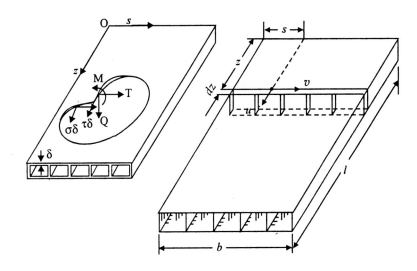

Fig. 3.1

The displacements of any point on the middle of the shell are represented in the form of the following resolutions:

$$u(z, s) = \sum_{1}^{m} U_i(z) \cdot \varphi_i(s), \quad (i = 1, 2, \ldots, m);$$

$$v(z, s) = \sum_{1}^{n} V_k(z) \cdot \psi_k(s), \quad (k = 1, 2, \ldots, n); \qquad \ldots (3.1)$$

in which $U_i(z)$ and $V_k(z)$ represent the unknown functions respectively of longitudinal and lateral displacements, while $\varphi_i(s)$ and $\psi_k(s)$ have to be selected beforehand.

In the terminology of V.Z. Vlasov:

$U_i(z)$ represent the generalised longitudinal displacements;

$\varphi_i(s)$ the generalized deformation co-ordinates of lateral elementary strips from the plane of the cross-section;

m the number of degrees of freedom of the elementary strip from the plane of the section equal to the number of nodes in the lateral section of the shell;

$V_k(z)$ the generalized lateral displacements;

$\psi_k(s)$ the generalized deformation co-ordinates of the lateral elementary strips in the plane of the cross-section; and

n the number of degrees of freedom of the section regarded as a hinged-pivotal system with inextensible members ($n = 2 \times m - c$, in which c is the number of rods in the elementary frame). The

m functions of U_i and n functions of V_k wholly determine the deformed state of the shell. By knowing it, the normal and shear stresses can be determined on the basis of Hooke's law at any point in the midsurface of the shell:

$$\sigma(z, s) = E \frac{\partial u}{\partial z} = E \cdot \sum_1^m U_i' (z) \cdot \varphi_i(s),$$

$$r(z, s) = G\left(\frac{\partial u}{\partial s} + \frac{\partial v}{\partial z}\right) = G\left[\sum_1^m U_i(z) \cdot \varphi_i'(s) + \sum_1^n V_k'(z) \cdot \psi_k(s)\right].$$

Differential equilibrium equations are drawn using the variational method for determining $U_i(z)$ and $V_k(z)$. For this purpose, the work of the forces of an elementary lateral strip on possible displacements represented by functions $\varphi_i(s)$ and $\psi_k(s)$ is examined. These equations have the following form:

$$\gamma \sum_1^m a_{ji} U_i'' - \sum_1^m b_{ji} U_i - \sum_1^m c_{jk} V_k' + \frac{1}{G} P_j = 0$$

$$\sum_1^n c_{hi} U_i' + \sum_1^n r_{hk} V_k'' - \gamma \sum_1^n s_{hk} V_k + \frac{1}{G} q_h = 0; \qquad \text{... (3.2)}$$

$$(j, i = 1, 2, \ldots, m), \ (h, k = 1, 2, \ldots, n).$$

Here

$$\gamma = \frac{E}{G}, \ a_{ji} = \int \varphi_j(s) \cdot \varphi_i(s) \cdot dF,$$

$$b_{ji} = \int \varphi_j'(s) \cdot \varphi_i'(s) \cdot dF, \ c_{jk} = \int \varphi_j'(s) \cdot \psi_k(s) \cdot dF, \ c_{hi} = \int \psi_h(s) \cdot \varphi_i'(s) \cdot dF,$$

$$r_{hk} = \int \psi_h(s) \cdot \varphi_k(s) \cdot dF, \ s_{hk} = \frac{1}{E} \int \frac{M_h \cdot M_k}{EJ} \cdot ds. \qquad \text{... (3.3)}$$

The integrals are extended to all the members of the cross-section, and $M_h(s)$ and $M_k(s)$ are determined by the structural mechanics methods by inversion from the deformed state of the strip to the internal forces.

The resultant system has $(m + n)$ orders and comprises $(m + n)$ linear differential ordinary equations with constant coefficients each of the second order relative to $(m + n)$ unknown functions $U_i(z)$ and $V_k(z)$.

Functions $U_i(z)$ determine the warping of the shell cross-section and functions $V_k(z)$ the deformation of the section contour.

The given functions $\varphi_i(s)$ and $\psi_k(s)$ should be linearly independent mutually and continuous at all points of the contour. Moreover, these functions should satisfy the main geometric hypotheses adopted for shells of medium length: the hypothesis of flat sections, for the strips constituting the shell and their lateral inextensibility. In other respects, these functions may differ.

The simplest form of functions $\varphi_i(s)$ and $\psi_k(s)$ adopted, for practical purposes, in all analyses of multiconnected prismatic shells of medium length, was proposed by V.Z. Vlasov (Fig. 3.2).

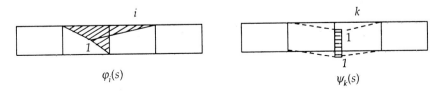

Fig. 3.2

3.2.2 Function of Contour Deformation as a Result of Static Method

Evidently, the method described by V.Z. Vlasov generates no error if one of the two groups of functions, viz., $\psi_k(s)$, is fixed by the static method, i.e., is in the form of displacement diagrams of a multiconnected strip of frame due to lateral loading similar to the given loading. Special advantages of such an approach are, however, possible in the case of a shell with a periodic structure since the single function $\psi(s)$ obtained in place of n functions of $\psi_k(s)$ assumes a closed analytical form. The single function can be called the *contour deformation function*; it greatly simplifies the calculation of coefficients c_{jk}, c_{hi}, r_{hk}, and especially s_{hk}.

A solution to the problem of flexure of a periodic multipanelled frame can be obtained by the method of forces as well as by the method of displacements. Since the solution for a shell by the method of V.Z. Vlasov comprises displacements, the auxiliary problem of finding the function $\psi(s)$ is best resolved by the displacements method.

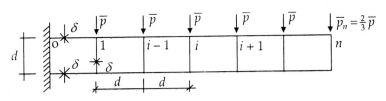

Fig. 3.3

Let the form of the cross-section of the prismatic shell be as shown in Fig. 3.3. Let us apply the displacements method in a developed form, adopting the rotation angles of nodes ω_i and their longitudinal displacements ψ_i as the main unknowns. Considering the casings as inextensible in the lateral direction, let us express the rotation angles of members within a given panel (cell) in terms of the vertical displacements of nodes:

$$\Omega_{i,\,i-1} = \frac{1}{d}\left(\psi_i - \psi_{i-1}\right),\ \Omega_{i,\,i+1} = \frac{1}{d}\left(\psi_{i+1} - \psi_i\right). \qquad \dots (3.4)$$

Taking into consideration eqn. (3.4), the equilibrium equation for node i will be:

$$M_{i,\,i+1} + M_{i,\,i-1} + M_{i,\,i} = 4\frac{EJ}{d}\cdot\omega_i + 2\frac{EJ}{d}\omega_{i+1} - 6\frac{EJ}{d}\Omega_{i,\,i+1} +$$

$$4\frac{EJ}{d}\omega_i + 2\frac{EJ}{d}\omega_{i-1} - 6\frac{EJ}{d}\Omega_{i,\,i-1} + 6\frac{EJ}{d}\omega_i = 0. \qquad \dots (3.5)$$

Rotation angles $\Omega_{i,\,i+1}$ and $\Omega_{i,\,i-1}$ are determined from the equilibrium conditions of the intersecting forces in cells i and $i-1$ respectively.

$$\Omega_{i,\,i+1} = \frac{1}{2}\left(\omega_i + \omega_{i+1}\right) + \frac{\overline{P}d^2}{24EJ}\left(n - i - \frac{1}{3}\right),$$

$$\Omega_{i,\,i-1} = \frac{1}{2}\left(\omega_i + \omega_{i-1}\right) + \frac{\overline{P}d^2}{24EJ}\left(n - i + \frac{2}{3}\right). \qquad \dots (3.6)$$

After substituting eqn. (3.6) in (3.5), the equilibrium equation for node i will be:

$$-\omega_{i-1} + 8\omega_i - \omega_{i+1} = \frac{\overline{P}\,d^2}{4EJ}\left(2n - 2i + \frac{1}{3}\right). \qquad \dots (3.7)$$

Equation (3.7) is regarded as a heterogeneous equation in finite differences. The general solution to the homogeneous equation corresponding to it, according to Bleikh and Melan (1938), will be:

$$\omega_i^0 = c_1\lambda^i + c_2\lambda^{-i}, \qquad \dots (3.8)$$

in which λ represents the root of the characteristic equation:

$$\lambda^2 - 8\lambda + 1 = 0. \qquad \dots (3.9)$$

As its roots are real ($\lambda = 7.872983$ and $1/\lambda = 0.127017$), the special solution is found in the form

$$\omega_i^{00} = (A + i\cdot B)\overline{P}.$$

After substituting the above in eqn. (3.7), we obtain:

$$A = \frac{d^2}{24EJ}\left(2n + \frac{1}{3}\right),$$

$$B = \frac{2d^2}{24EJ}.$$

The complete solution to eqn. (3.7) will have the following form:

$$\omega_i = c_1\,\lambda^i + c_2\,\lambda^{-1} + \frac{\overline{P}\,d^2}{24EJ}\left(2n - 2i + \frac{1}{3}\right). \qquad \dots (3.10)$$

Constants C_1 and C_2 are determined from the limiting conditions:

$$\omega_0 = 0, \qquad \dots (3.11)$$

$$7\omega_n - \omega_{n-1} = \frac{\overline{P}\,d^2}{6EJ}. \qquad \dots (3.12)$$

Satisfying these conditions, we find:

$$C_1 = \frac{\overline{P}\,d^2}{24EJ}\left(2n + \frac{1}{3}\right)\Gamma, \quad C_2 = -\frac{\overline{P}\,d^2}{24EJ}\left(2n + \frac{1}{3}\right)(1+\Gamma),$$

in which

$$\Gamma = \frac{\dfrac{12}{6n+1} + (7-\lambda)\lambda^{-n}}{\left(7-\dfrac{1}{\lambda}\right)\lambda^n - (7-\lambda)\lambda^{-n}} = \frac{\dfrac{13.745972}{6n+1} - \lambda^{-n}}{7.868358\lambda^n + \lambda^{-n}}.$$

Then, eqn. (3.10) assumes the form:

$$\omega_i = \frac{\overline{P}\,d^2}{24EJ}\left(2n + \frac{1}{3}\right)\left[1 - \frac{2i}{2n+\dfrac{1}{3}} - \frac{1}{\lambda^i} + \left(\lambda^i - \frac{1}{\lambda_i}\right)\cdot\Gamma\right].$$

From kinematic concepts, $\psi_i = d\sum_{1}^{i}\Omega_{i,i-1}$.

Substituting in it eqns. (3.6) and (3.12) and summing up, we derive

$$\psi_i = \frac{\overline{P}\,d^3}{48EJ}\left(2n + \frac{1}{3}\right)\left\{3i - \frac{3i^2}{2n+\dfrac{1}{3}} - \frac{\lambda+1}{\lambda-1}\left(1 - \frac{1}{\lambda^i}\right)\left[1 - (\lambda^i - 1)\cdot\Gamma\right]\right\}.$$

$$\dots (3.13)$$

Finally, the known function of contour deformation acquires the form:

$$\psi_i(s) = \psi_{i-1} + \omega_{i-1}\cdot s + M_{i-1,i}\cdot\frac{s^2}{2} + Q_{i-1,i}\cdot\frac{s^3}{6} =$$

$$\psi_{i-1}+(\psi_i-\psi_{i-1})\frac{s}{d}+A_i\cdot\frac{s}{2}\left(\frac{s}{d}-1\right)+B_i\cdot\frac{sd}{6}\left(\frac{s^2}{d^2}-1\right), \qquad \text{.... (3.14)}$$

in which

$$A_i=\frac{\overline{P}d^2}{48EJ}\left(2n+\frac{1}{3}\right)\left\{\left[-3+\frac{6i-1}{2n+\frac{1}{3}}-\frac{2}{\lambda^i}(1+2\lambda)+2\left[\lambda^i\left(1+\frac{2}{\lambda}\right)-\frac{1}{\lambda^{i-1}}(1+2\lambda)\right]\cdot\Gamma-\right.\right.$$

$$\left.\left.3\frac{\lambda+1}{\lambda-1}\left\{\left(1-\lambda^{1-i}\right)\left[1-\left(\lambda^{i-1}-1\right)\cdot\Gamma\right]-\left(1-\lambda^{-i}\right)\left[1-\left(\lambda^i-1\right)\cdot\Gamma\right]\right\}\right]\right\}$$

$$B_i=\frac{\overline{P}d}{4EJ}\left(2n+\frac{1}{3}\right)\left\{\left\{1-\frac{2i-1}{2n+\frac{1}{3}}+\frac{\lambda+1}{\lambda^i}-\Gamma\left[\lambda^i\left(1+\frac{1}{\lambda}\right)+\lambda^{-i}(1+\lambda)\right]+\right.\right.$$

$$\left.\left.\frac{\lambda+1}{\lambda-1}\left\{\left(1-\lambda^{i-1}\right)\left[1-\left(\lambda^{i-1}-1\right)\cdot\Gamma\right]-\left(1-\lambda^{-i}\right)\left[1-\left(\lambda^i-1\right)\cdot\Gamma\right]\right\}\right\}\right\}.$$

Figure 3.4 shows the displacement diagram $\psi(s)$ constructed using eqn. (3.14).

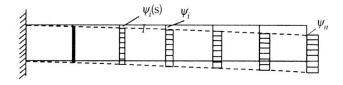

Fig. 3.4

The derived function of contour deformation satisfies the condition of continuity at all points of the shell cross-section. Moreover, it automatically satisfies the limiting conditions on the longitudinal edges of the shell.

3.2.3 Reduction of the Order of the System of Equations by Introducing a Single Function of Section Contour Deformation

By introducing a single function of contour deformation $\psi(s)$ in the second part of eqn. (3.1), only one member remains:

$$v(z,s)=V(z)\cdot\psi(s), \qquad \text{... (3.15)}$$

in which $V(z)$ becomes the function of generalised displacements substituting all values of $V_k(z)$: $V_1(z), V_2(z),...,V_n(z)$.

Then, only one equation remains in the second group of variational equations of equilibrium (3.2) and the entire system (3.2) assumes a simpler form:

$$\gamma \sum_1^m a_{ji} U_i'' - \sum_1^m b_{ji} U_i - c_{i0} V' + \frac{1}{G} P_j = 0,$$

$$c_{0i} U_i' + r_0 V'' - \gamma s_0 V + \frac{1}{G} q_0 = 0; \qquad \ldots (3.16)$$

$$(i, j = 1, 2, \ldots, m)$$

Here:

$$a_{ji} = \int \varphi_{ji}(s) \cdot \varphi_i(s) \cdot dF, \ b_{ji} = \int \varphi_j'(s) \cdot \varphi_j' \cdot dF, \ c_{i0} = c_{0i} = \qquad \ldots (3.17)$$

$$\int \varphi_i'(s) \cdot \psi(s) \cdot dF, \ r_0 = \int \psi^2(s) \cdot dF, \ s_0 = \iint \left[\psi''(s) \right]^2 \cdot ds.$$

The system of equations (3.16) is of order $(m + 1)$. Thus, by introducing a single function $\psi(s)$ instead of m functions of $\psi_k(s)$, the order of the system of equations can be reduced from $(m + n)$ to $(m + 1)$.

The maximum reduction of the order of the system of equations is twice in shells with edges of uniform thickness. This is of practical interest even when using computer techniques. More importantly, however, simplification of the mathematical aspect of the problem for such shells helps solve very complex problems, e.g., analysing the impact of mobile loadings.

3.2.4 *Reducing the Labour of Computing the Coefficients of Equations*

Apart from reducing the order of the system of equations of V.Z. Vlasov's method, calculations for the coefficients $s_{hk} = \frac{1}{E} \int \frac{M_h \cdot M_k}{EJ} ds$ are considerably reduced by introducing the single function $\psi(s)$. In the conventional method for shells with m cells in the cross-section, m systems of algebraic equations of the m-th order with m unknowns have to be drawn and solved for determining $M_h(s)$ and $M_k(s)$ and later "multiplied" $(m/2)$ by $(m+1)$ times M_h and M_k for calculating all the coefficients of s_{hk}.

Using function $\psi(s)$, only one coefficient $s_0 = \iint \left[\psi''(s) \right]^2 \cdot ds$ is determined for any number of cells in the cross-section of the shell.

3.2.5 *Application of Theory to Analysing Shells of Medium Length with Periodic Structure*

To illustrate and compare with the conventional method, let us analyse a prismatic shell with multiconnected section comprising six identical cells (Fig. 3.5).

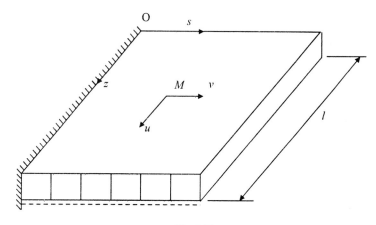

Fig. 3.5

One of the longitudinal edges of the shell is firmly fixed, the opposite edge is free, while the shell is hinge-supported on the lateral edges. The weight of the shell itself reduced to the nodal lines has been adopted as the load. As in Vlasov (1958), let us assume that the shell is made of reinforced concrete with the following parameters:

- number of cells in the section, $m = 6$,
- size of panel, $d_1 = d = 6$ m,
- height of walls (ribs), $d_2 = d = 6$ m,
- thickness of casing and wall $\delta_1 = \delta_2 = \delta = 0.08$ m,
- density of the material, 2.4×10^3 kg/m^3,
- modulus of elasticity, $E = 2 \times 10^4$ MPa,
- $\gamma = E/G = 2$,
- weight of the shell reduced to any intermediate node of strip of frame of width $dz = 3 \times 2.4 \times 10^3 \times d \times \delta = 3 \times 1.152 \times 10^3$ kg/m corresponding to loading $3 \times p = 3 \times 1.152 \times 10^4$ N/m, and
- length of shell, $l = 60$ m.

A cross-section of the above shell is shown in Figure 3.6.

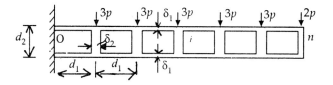

Fig. 3.6

(a) Analysis by the conventional method

Taking into account the shell symmetry relative to the common midsurface: $m = 6$ and $n = 6$.

Displacement of any point on the midsurface of shell is represented in the form of a finite series:

$$u(z,s) = U_1(z)\cdot\varphi_1(s)+U_2(z)\cdot\varphi_2(s)+U_3(z)\cdot\varphi_3(s)+$$

$$U_4(z)\cdot\varphi_4(s)+U_5(z)\cdot\varphi_5(s)+U_6(z)\cdot\varphi_6(s), \qquad \dots (3.18)$$

$$v(z,s) = V_1(z)\cdot\psi_1(s)+V_2(z)\cdot\psi_2(s)+V_3(z)\cdot\psi_3(s)+$$

$$V_4(z)\cdot\psi_4(s)+V_5(z)\cdot\psi_5(s)+V_6(z)\cdot\psi_6(s).$$

Functions $\varphi_i(s)$ and $\psi_k(s)$ figuring in eqn. (3.18) are shown in the form of diagrams in Fig. 3.7.

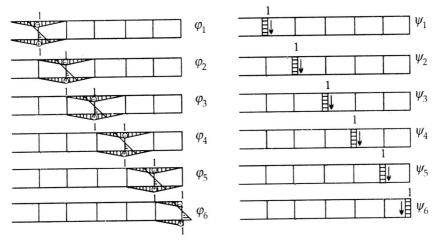

Fig. 3.7

The diagrams of the derivatives of functions $\varphi_i(s)$ are shown in Fig. 3.8. The system of differential equations (3.2) relative to the unknown displacements of U_i and V_k is shown as a matrix (Table 3.1).

The coefficients of equations (3.2) are calculated using equation (3.3) based on the diagrams of functions ψ_i, ψ_k, φ_i' (Figs. 3.7 and 3.8):

$$a_{11} = a_{22} = a_{33} = a_{44} = a_{55} = \frac{5}{3}d\delta, \; a_{66} = d\delta,$$

$$a_{12} = a_{23} = a_{34} = a_{45} = a_{56} = \frac{1}{3}d\delta,$$

$$b_{11} = b_{22} = b_{33} = b_{44} = b_{55} = 8\frac{\delta}{d}, \; b_{66} = 6\frac{\delta}{d}, \qquad \dots (3.19)$$

$$b_{12} = b_{23} = b_{34} = b_{45} = b_{56} = -2\frac{\delta}{d},$$

$$c_{11} = c_{22} = c_{33} = c_{44} = c_{55} = c_{66} = 2d,$$

$$r_{11} = r_{22} = r_{33} = r_{44} = r_{55} = r_{66} = d\delta.$$

Table 3.1

	$U_1(z)$	$U_2(z)$	$U_3(z)$	$U'_4(z)$	$U_5(z)$	$U_6(z)$	$V_1(z)$	$V_2(z)$	$V_3(z)$	$V_4(z)$	$V_5(z)$	$V_6(z)$	Free member
1	A_{11}	A_{12}	0	0	0	0	$-c_{11}D_2$	0	0	0	0	0	0
2	A_{21}	A_{22}	A_{23}	0	0	0	0	$-c_{22}D_2$	0	0	0	0	0
3	0	A_{32}	A_{33}	A_{34}	0	0	0	0	$-c_{33}D_2$	0	0	0	0
4	0	0	A_{43}	A_{44}	A_{45}	0	0	0	0	$-c_{44}D_2$	0	0	0
5	0	0	0	A_{54}	A_{55}	A_{56}	0	0	0	0	$-c_{55}D_2$	0	0
6	0	0	0	0	A_{65}	A_{66}	0	0	0	0	0	$-c_{66}D_2$	0
7	$c_{11}D_1$	0	0	0	0	0	R_{11}	$-\gamma s_{12}$	$-\gamma s_{13}$	$-\gamma s_{14}$	$-\gamma s_{15}$	$-\gamma s_{16}$	q_1/G
8	0	$c_{22}D_1$	0	0	0	0	$-\gamma s_{21}$	R_{22}	$-\gamma s_{23}$	$-\gamma s_{24}$	$-\gamma s_{25}$	$-\gamma s_{26}$	q_2/G
9	0	0	$c_{33}D_1$	0	0	0	$-\gamma s_{31}$	$-\gamma s_{32}$	R_{33}	$-\gamma s_{34}$	$-\gamma s_{35}$	$-\gamma s_{36}$	q_3/G
10	0	0	0	$c_{44}D_1$	0	0	$-\gamma s_{41}$	$-\gamma s_{42}$	$-\gamma s_{43}$	R_{44}	$-\gamma s_{45}$	$-\gamma s_{46}$	q_4/G
11	0	0	0	0	$c_{55}D_1$	0	$-\gamma s_{51}$	$-\gamma s_{52}$	$-\gamma s_{53}$	$-\gamma s_{54}$	R_{55}	$-\gamma s_{56}$	q_5/G
12	0	0	0	0	0	$c_{66}D_1$	$-\gamma s_{61}$	$-\gamma s_{62}$	$-\gamma s_{63}$	$-\gamma s_{64}$	$-\gamma s_{65}$	R_{66}	q_6/G

$$D_1 = -\frac{\pi}{e}, \quad D_1^2 = -\frac{\pi^2}{e^2}, \quad D_2 = \frac{\pi}{e}, \quad D_2^2 = -\frac{\pi^2}{e^2}$$

$A_{11} = \gamma a_{11} D_1^2 - b_{11}$, $A_{12} = \gamma a_{12} D_1^2 - b_{12}$,

$A_{22} = \gamma a_{22} D_1^2 - b_{22}$, $A_{21} = \gamma a_{21} D_1^2 - b_{21}$,

$A_{33} = \gamma a_{33} D_1^2 - b_{33}$, $A_{23} = \gamma a_{23} D_1^2 - b_{23}$, $A_{45} = \gamma a_{45} D_1^2 - b_{45}$,

$A_{44} = \gamma a_{44} D_1^2 - b_{44}$, $A_{32} = \gamma a_{32} D_1^2 - b_{32}$, $A_{54} = \gamma a_{54} D_1^2 - b_{54}$,

$A_{55} = \gamma a_{55} D_1^2 - b_{55}$, $A_{34} = \gamma a_{34} D_1^2 - b_{34}$, $A_{56} = \gamma a_{56} D_1^2 - b_{56}$,

$A_{66} = \gamma a_{66} D_1^2 - b_{66}$, $A_{43} = \gamma a_{43} D_1^2 - b_{43}$, $A_{65} = \gamma a_{65} D_1^2 - b_{65}$.

$R_{11} = r_{11} D_2^2 - \gamma s_{11}$,

$R_{22} = r_{22} D_2^2 - \gamma s_{22}$,

$R_{33} = r_{33} D_2^2 - \gamma s_{33}$,

$R_{44} = r_{44} D_2^2 - \gamma s_{44}$,

$R_{55} = r_{55} D_2^2 - \gamma s_{55}$,

$R_{66} = r_{66} D_2^2 - \gamma s_{66}$.

To calculate the coefficients s_{hk} by backward sweep from the deformed state of the strip (Fig. 3.6) to the internal forces, six systems of equations are drawn and solved relative to the nodal moments by the method of displacements of the sixth order each. The main system of the method of displacements constructed using symmetry is shown in Fig. 3.9 in which are given 7, 8, 9, 10, 11 and 12—additional nodes preventing the longitudinal displacements of nodes 1, 2, 3, 4, 5, 6 respectively; $i_1 = E_1 J_1 / d_1$ and $i_2 = E_2 J_2 / d_2$.

The equations of the displacement method can be formulated in a developed form (Kornoukhov, 1949). Then, the equilibrium equations of nodes for the basic system of equations under consideration will have the following form:

$$M_{10} + M_{12} + M_{17} = 0, \qquad M_{21} + M_{23} + M_{28} = 0,$$

$$M_{32} + M_{34} + M_{39} = 0, \qquad M_{43} + M_{45} + M_{4,10} = 0, \qquad \text{... (3.20)}$$

$$M_{54} + M_{56} + M_{5,11} = 0, \qquad M_{65} + M_{6,12} = 0.$$

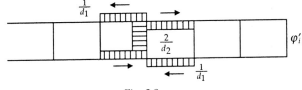

Fig. 3.8

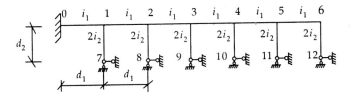

Fig. 3.9

By expressing nodal moments in terms of boundary displacements of rods, we find

$$-i_1 \left(4\omega_1 + 2\omega_0 - 6\theta_{10}\right) - i_1 \left(4\omega_1 + 2\omega_2 - 6\theta_{12}\right) - 2i_2 \left(3\omega_1 - 3\theta_{17}\right) = 0,$$

$$-i_1 \left(4\omega_2 + 2\omega_1 - 6\theta_{21}\right) - i_1 \left(4\omega_2 + 2\omega_3 - 6\theta_{23}\right) - 2i_2 \left(3\omega_2 - 3\theta_{28}\right) = 0,$$

$$-i_1 \left(4\omega_3 + 2\omega_2 - 6\theta_{32}\right) - i_1 \left(4\omega_3 + 2\omega_4 - 6\theta_{34}\right) - 2i_2 \left(3\omega_3 - 3\theta_{39}\right) = 0,$$

$$-i_1 \left(4\omega_4 + 2\omega_3 - 6\theta_{43}\right) - i_1 \left(4\omega_4 + 2\omega_5 - 6\theta_{45}\right) - 2i_2 \left(3\omega_4 - 3\theta_{4,10}\right) = 0,$$

$$-i_1 \left(4\omega_5 + 2\omega_4 - 6\theta_{54}\right) - i_1 \left(4\omega_5 + 2\omega_6 - 6\theta_{56}\right) - 2i_2 \left(3\omega_5 - 3\theta_{5,11}\right) = 0,$$

$$-i_1 \left(4\omega_6 + 2\omega_5 - 6\theta_{65}\right) - 2i_2 \left(3\omega_6 - 3\theta_{6,12}\right) = 0. \qquad \text{... (3.21)}$$

Here, ω is the rotation angle of nodes and θ the rotation angle of rods. Let us determine the nodal moments for $\psi_1 = 1$.

In this case, $\theta_{01} = \dfrac{1}{d_1}$, $\theta_{12} = \theta_{21} = -\dfrac{1}{d_1}$; all the rest of the rotation angles are equal to zero. By substituting these values in the system of eqns. (3.21) and solving it by the Gauss method, we find the angles of rotation of nodes ω_1, ω_2, ω_3, ω_4, ω_5 and ω_6. Later, the nodal moments M_{01}, M_{10}, M_{12}, $\ldots$, $M_{6.12}$ are calculated and from them the diagram of moments $\overline{M}_1$ is constructed for $\psi_1 = 1$.

Diagram $\overline{M}_2$ is constructed similarly. Let us assume that $\psi_2 = 1$; then $\theta_{21} = \theta_{12} = 1/d_1$ and $\theta_{23} = \theta_{32} = -1/d_1$, all other θ angles being equal to zero. By resolving the systems of equations (3.21) for these initial data, we find the rotation angles of nodes and later the nodal moments M_{01}, M_{10}, M_{12},...., $M_{6.12}$ and construct from them diagram $\overline{M}_2$. For constructing figure $\overline{M}_3$, $\psi_3 = 1$, $\theta_{23} = \theta_{32} = -1/d_1$, $\theta_{34} = \theta_{43} = -1/d_1$ should be plotted, all other θ values being equal to zero. For constructing figure $\overline{M}_4$, it is assumed that $\psi_4 = 1$, $\theta_{34} = \theta_{43} = -1/d_1$, $\theta_{45} = \theta_{54} = -1/d_1$. For constructing figure $\overline{M}_5$: $\psi_5 = 1$, $\theta_{45} = \theta_{54} = -1/d_1$; $\theta_{56} = \theta_{65} = -1/d_1$. Each of the figures of moments constructed in this manner corresponds to the given function $\psi_k(s)$ (Fig. 3.7). With these diagrams and "multiplying" them in conformity with the last one of eqn. (3.3), we find 21 coefficients of s_{hk}.

Let diaphragms be present on shell ends, absolutely rigid in the plane and flexible across the plane of the section. Then, the limiting conditions corresponding to the absence of transverse displacements and normal stresses at the shell ends at $z = 0$ and at $z = 1$ will be: $U_i'(z) = 0$ and $V_k(z) = 0$. Using the method of trigonometric series for integrating the system of equations (3.21) and restricting to the first member of the resolution, we obtain the following equations for the unknown functions $U_i(z)$, and $V_k(z)$.

$$U_i(z) = U_i \cos\frac{\pi z}{e}, \quad V_k(z) = V_k \sin\frac{\pi z}{e}, \quad q_k(z) = \frac{4}{\pi} q_k \sin\frac{\pi z}{e} \ \ldots \ (3.22)$$

in which

$$q_k = 3 \cdot p \ (k = 1, 2, 3, 4, 5), \ q_6 = 2 \cdot p.$$

After substituting eqn. (3.22) in eqns. (3.21) and reducing to $\cos\pi\,z/e$ and $\sin\pi\,z/e$, we obtain a system of algebraic equations relative to the unknown coefficients U_i and V_k (Table 3.2).

Table 3.2

No. of Node	1	2	3	4	5	6
v_i	12543	18743	22747	25541	27379	28700
u_i	1524	2488	3104	3551	3764	3983

For the initial data adopted in the example under consideration, the terms of the matrix of equations (Table 3.1) will have the following values:

$$A_{11} = \gamma a_{11} D_1^2 - b_{11} = -111053 \cdot 10^{-6}, \quad A_{22} = A_{33} = A_{44} = A_{55} = A_{11},$$

$$A_{66} = -82632 \cdot 10^{-6}, \quad A_{12} = \gamma a_{12} D_1^2 - b_{12} = 25789 \cdot 10^{-6},$$

$$A_{23} = A_{34} = A_{45} = A_{56} = A_{12}, \quad A_{32} = A_{43} = A_{54} = A_{65} = A_{12},$$

$$R_{11} = r_{11} D_2^2 - \gamma s_{11} = -1330, 8 \cdot 10^{-6}, \quad R_{66} = -1321, 2 \cdot 10^{-6},$$

$$R_{22} = R_{33} = R_{44} = R_{55} = -1330.8 \cdot 10^{-6},$$

$$-\gamma s_{12} = -\gamma s_{21} = 9.4646 \cdot 10^{-6}, \qquad -\gamma s_{13} = -\gamma s_{31} = 2.07788 \cdot 10^{-6},$$

$$-\gamma s_{14} = -\gamma s_{41} = 0.030474 \cdot 10^{-6}, \qquad -\gamma s_{15} = -\gamma s_{51} = -0.046105 \cdot 10^{-6},$$

$$-\gamma s_{16} = -\gamma s_{61} = -0.0051915 \cdot 10^{-6}, \quad -\gamma s_{23} = -\gamma s_{32} = -9.21038 \cdot 10^{-6},$$

$$-\gamma s_{24} = -\gamma s_{42} = -1.64406 \cdot 10^{-6}, \qquad -\gamma s_{25} = -\gamma s_{52} = 0.29874 \cdot 10^{-6},$$

$$-\gamma s_{26} = -\gamma s_{62} = 0.042998 \cdot 10^{-6}, \qquad -\gamma s_{34} = -\gamma s_{43} = 8.5342 \cdot 10^{-6},$$

$$-\gamma s_{35} = -\gamma s_{53} = -2.01433 \cdot 10^{-6}, \qquad -\gamma s_{36} = -\gamma s_{63} = 0.24434 \cdot 10^{-6},$$

$$-\gamma s_{45} = -\gamma s_{54} = 9.75304 \cdot 10^{-6}, \qquad -\gamma s_{46} = -\gamma s_{64} = 1.67349 \cdot 10^{-6},$$

$$-\gamma s_{56} = -\gamma s_{65} = 6.72954 \cdot 10^{-6},$$

$$q_1 = q_2 = q_3 = q_4 = q_5 = 3.81972 \, p, \quad q_6 = 2.54648 \, p.$$

The solutions derived are shown in Table 3.2.

These same results are represented in the form of displacement diagrams of

–longitudinal in the support section $z = 0$ (diagram u in Fig. 3.10, continuous line) and

–transverse in the midsection $z = 1/2$ (diagram v in Fig. 3.11, continuous line).

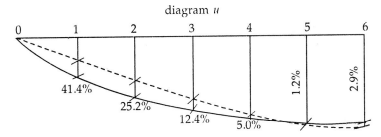

diagram u

Fig. 3.10

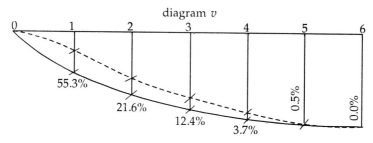

Fig. 3.11

b) *Analysis using the periodicity of shell structure*

In this case, the displacements of an arbitrary point on the midsurface of the shell under consideration are represented by expressions:

$$u(z,s) = U_1(z)\cdot\varphi_1(s) + U_2(z)\cdot\varphi_2(s) + U_3(z)\cdot\varphi_3(s) +$$

$$U_4(z)\cdot\varphi_4(s) + U_5(z)\cdot\varphi_5(s) + U_6(z)\cdot\varphi_6(s), \qquad \dots (3.23)$$

$$u(z,s) = V(z)\cdot\psi(s),$$

in which functions $\varphi_i(s)$ are adopted in the form shown in Fig. 3.7 and the contour deformation function $\psi(s)$ according to eqn. (3.14) at $P = 1$ (the corresponding diagram is shown in Fig. 3.4). The system of differential equations (3.16) of order $(m + 1) = (6 + 1) = 7$ is given in matrix form in Table 3.3. The equation coefficients are calculated according to eqns. (3.17). Evidently, a_{ji} and b_{ji} have the former values and coefficients:

$$c_{0i} = c_{i0} = \int \psi_i \cdot \psi_i'\cdot(s)\cdot dF = d\cdot\delta\cdot\psi_i\cdot\psi_i'\cdot(s), \quad r_0 = \int \psi_i^2 \cdot dF = d\cdot\delta\cdot\sum_1^n \psi_i^2,$$

$$s_0 = \int\int \left[\psi''(s)\right]^2 \cdot ds = 4J\left[2d\sum_1^d N_i^2 + \frac{3}{d}\sum_1^n \Phi_i^2\right], \qquad \dots (3.24)$$

$$N_i = -\frac{\overline{P}d}{48EJ}\left(2n+\frac{1}{3}\right)\left[\frac{3i-0.5}{2n+\frac{1}{3}} + \frac{1-\lambda}{2\lambda^i}(1+\Gamma)-\frac{3}{2}\right],$$

$$\Phi_i = \frac{\overline{P}d^2}{48EJ}\left(2n+\frac{1}{3}\right)\left[1 - \frac{2i}{2n+\frac{1}{3}} - \frac{1}{\lambda^i} + \left(\lambda^i - \frac{1}{\lambda^i}\right)\cdot\Gamma\right].$$

The free term $q_0 = \sum_1^n q_i \times \psi_i$. The same method may be adopted for integration. For the unknown functions and the free term, we have

$$U_i(z) = U_i\cos\frac{\pi z}{l}, \quad V(z) = V\cdot\sin\frac{\pi z}{l}, \quad q_0 = \frac{4}{\pi}p\left(3\sum_1^5 \psi_i + 2\psi_6\right)\cdot\sin\frac{\pi z}{l} \quad \dots (3.25)$$

Table 3.3

	$U_1(z)$	$U_2(z)$	$U_3(z)$	$U_4(z)$	$U_5(z)$	$U_6(z)$	$V(z)$	Free member
1	$\gamma a_{11}D_1^2 - b_{11}$	$\gamma a_{12}D_1^2 - b_{12}$	0	0	0	0	$-c_{01}D_2$	0
2	$\gamma a_{21}D_1^2 - b_{21}$	$\gamma a_{22}D_1^2 - b_{22}$	$\gamma a_{23}D_1^2 - b_{23}$	0	0	0	$-c_{02}D_2$	0
3	0	$\gamma a_{32}D_1^2 - b_{32}$	$\gamma a_{33}D_1^2 - b_{33}$	$\gamma a_{34}D_1^2 - b_{34}$	0	0	$-c_{03}D_2$	0
4	0	0	$\gamma a_{43}D_1^2 - b_{43}$	$\gamma a_{44}D_1^2 - b_{44}$	$\gamma a_{45}D_1^2 - b_{45}$	0	$-c_{04}D_2$	0
5	0	0	0	$\gamma a_{54}D_1^2 - b_{54}$	$\gamma a_{55}D_1^2 - b_{55}$	$\gamma a_{56}D_1^2 - b_{56}$	$-c_{05}D_2$	0
6	0	0	0	0	$\gamma a_{65}D_1^2 - b_{65}$	$\gamma a_{66}D_1^2 - b_{66}$	$-c_{06}D_2$	0
7	$c_{10}D_1$	$c_{20}D_1$	$c_{30}D_1$	$c_{40}D_1$	$c_{50}D_1$	$c_{60}D_1$	$r_0 D_2^2 - \gamma s_0$	q_0/G

$c_{10}D_1 = -c_{01}D_2 = 0.92232 \cdot 10^{-6}$, $c_{20}D_1 = -c_{02}D_2 = -2.04699 \cdot 10^{-6}$, $c_{30}D_1 = -c_{03}D_2 = 3.009 \cdot 10^{-6}$, $c_{40}D_1 = -c_{04}D_2 = -3.7146 \cdot 10^{-6}$,

$c_{50}D_1 = -c_{05}D_2 = -4.1554 \cdot 10^{-6}$, $c_{60}D_1 = -c_{06}D_2 = -4.3329 \cdot 10^{-6}$, $r_0 D_2^2 - \gamma s_0 = -0.0012 \cdot 10^{-6}$, $q_0 = 76161 \cdot 10^{-6} p$.

Table 3.4

No. of node	1	2	3	4	5	6
v_i	5600	13553	19923	24595	27519	28689
u_i	893	1861	2718	3371	3810	4098

The matrix elements while maintaining the first term of the series have the following values:

$$\gamma a_{11} D_1^2 - b_{11} = \gamma a_{22} D_1^2 - b_{22} = \gamma a_{33} D_1^2 - b_{33} = \gamma a_{44} D_1^2 - b_{44} =$$
$$\gamma a_{55} D_1^2 - b_{55} = -111053 \cdot 10^{-6},$$
$$\gamma a_{12} D_1^2 - b_{12} = \gamma a_{23} D_1^2 - b_{23} = \gamma a_{34} D_1^2 - b_{34} = \gamma a_{45} D_1^2 - b_{45} =$$
$$\gamma a_{56} D_1^2 - b_{56} = -25789 \cdot 10^{-6},$$
$$\gamma a_{66} D_1^2 - b_{66} = -82632 \cdot 10^{-6},$$
$$c_{10} D_1 = -c_{01} D_2 = -0.92232 \cdot 10^{-6}, \quad c_{20} D_1 = -c_{02} D_2 = -2.04699 \cdot 10^{-6},$$
$$c_{30} D_1 = -c_{03} D_2 = -3.009 \cdot 10^{-6}, \quad c_{40} D_1 = -c_{04} D_2 = -3.7146 \cdot 10^{-6}.$$

c) Comparison of results

Numerical comparisons show that the results in the zone of maximum displacements nearly coincide. Differences prevail close to the fixed edge of the shell. An examination of displacement diagrams given in Figures 3.10 and 3.11 shows, however, that the static method of approximation corresponds best to the actual nature of fixing the longitudinal margins of the shell. Solutions are shown in Table 3.4 and diagrams for u (Fig. 3.10) and v (Fig. 3.11) as broken lines. The deviations are in percentage.

3.3 Analysis of Prismatic Shells of Multiconnected Section with Periodic Structure Ignoring Shear Deformation

3.3.1 Equations for V.Z. Vlaslov's Variational Displacements Method for Prismatic Shells of Multiconnected Section Ignoring Shear Deformation

These equations were derived by I.E. Mileikovskii (1950, b). His discussions and conclusions are briefly reviewed below.

Let us express the shear deformation in a strip using eqn. (3.1) and bring the equilibrium to zero:

$$\gamma = \frac{\partial u(z,s)}{\partial s} + \frac{\partial v(z,s)}{\partial z} = \sum_{l}^{m} U_i(z) \cdot \varphi_i'(s) + \sum_{l}^{n} V_k'(z) \cdot \psi_k(s) = 0$$

This condition can be satisfied by adopting $m = n$ and $i = k$.

Then, we find for the i-th state:

$$U_i(z) \cdot \varphi_i'(s) = -V_i'(z) \cdot \psi_i(s), \quad (i = 1, 2, \ldots, m) \qquad \ldots (3.26)$$

The above equation leads to the conclusion that the longitudinal deformation of the system wholly determines the lateral deformation and vice versa (Mileikovskii, 1950, a). If functions $\varphi_i'(s)$ and $\psi_i(s)$ are selected such that

$$\varphi_i'(s) = \psi_i(s), \qquad \ldots (3.27)$$

then

$$U_i(z) = -V_i'(z). \qquad \ldots (3.28)$$

The integral variational condition of equilibrium of an elementary strip in plane $z = $ const. has the following form:

$$\int \frac{\partial \tau}{\partial z} \psi_i(s) dF - \sum_i V_k(z) \int \frac{M_j(s) \cdot M_i(s)}{EJ} ds +$$

$$\int q(z) \cdot \psi_i(s) \cdot ds = 0, \quad (i, j = 1, \ldots . m). \qquad \ldots (3.29)$$

From the equilibrium condition of element $dz \times ds$ in direction z, it follows that

$$\frac{\partial}{\partial z}(\sigma \cdot \delta) + \frac{\partial}{\partial s}(\tau \cdot \delta) + p(z) = 0 \qquad \ldots (3.30)$$

Later, by integrating in parts the first term of eqn. (3.29) and substituting therein eqn. (3.30), using Hooke's law for normal stresses $\delta(z,s) = E\frac{\partial u}{\partial s} = E\sum U_i'(z) \cdot \varphi_i(s)$, and eqns. (3.27) and (3.28), and carrying out some conversions, we derive a system of differential equations for the strip relative to $V_i(z)$ in the following form:

$$\sum_i a_{ji} V_i^{1V}(z) + \sum_i s_{ji} V_i(z) - \frac{1}{E} q_j(z) - \frac{1}{E} p_j'(z) = 0, \qquad \ldots (3.31)$$

in which

$$a_{ji} = \int \varphi_j(s) \cdot \varphi_i(s) \cdot dF, \quad q_j(z) = \int q(z) \cdot \psi_j(s) \cdot ds,$$

$$s_{ji} = \frac{1}{E} \int \frac{M_j(s) \cdot M_i(s)}{EJ} ds, \quad p_j'(z) = \int p'(z) \cdot \varphi_j(s) \cdot ds,$$

$q(z)$ is the intensity of given lateral loading and
$p(z)$ the intensity of given longitudinal loading.

I.E. Mileikovskii derived the system of equations (3.31) in a more general form later (1960). The latter work demonstrated that equations (3.31) are also relevant for prismatic shells with an open as well as closed section, single or multiconnected.

Thus, by ignoring shear deformation, the problem of flexure of prismatic shells of a multiconnected section involves resolving a system of differential equations of order n.

3.3.2 Reduction of the Problem for Shells of Periodic Structure to a Single Differential Equation

For shells in which the influence of shear deformations is negligibly small, the mathematical aspect of the problem can be further simplified. If only one function of contour deformation $\psi(s)$ describing the contour deformability of a shell cross-section is fixed, the lone function of section warping can be determined on the basis of eqn. (3.27):

$$\varphi(s) = \int \psi(s) \cdot ds . \qquad \ldots (3.32)$$

Then, eqn. (3.31) is reduced to a single differential equation:

$$a_0 V^{1V}(z) + s_0 V(z) - \frac{1}{E} q_0(z) - \frac{1}{E} p_0'(z) = 0, \qquad \ldots (3.33)$$

in which

$$a_0 = \int [\varphi(s)]^2 \cdot dF = \int \left[\int \psi(s) \cdot ds \right]^2 \cdot dF, \quad s_0 = \int \int [\psi''(s)]^2 \cdot ds,$$

$$q_0(z) = \int q(z) \cdot \psi(s) \cdot ds, \quad p_0'(z) = \int p'(z) \cdot \varphi(s) \cdot ds. \qquad \ldots (3.34)$$

3.3.3 Application of Theory to Analysing Shells with a Multiconnected Section

Let us solve here the problem studied in Section 3.1.5, ignoring shear deformation in the strips forming the shell. The contour deformation function $\psi(s)$ is adopted in the form of displacement diagrams of a multiconnected strip of frame on auxiliary loading. This diagram for the shell under consideration is shown in Fig. 3.4. Then, the $\varphi(s)$ diagram according to eqn. (3.32) assumes the form shown in Fig. 3.12.

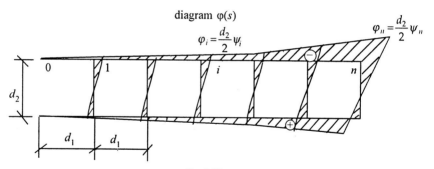

diagram $\varphi(s)$

Fig 3.12

By "multiplying" the $\varphi(s)$ diagram by itself as shown in eqn. (3.34) at $n = 6$, $d_1 = d_2 = d$ and $\delta_1 = \delta_2 = \delta$, we derive:

$$a_0 = \frac{5}{12} d^3 \delta \left\{ \sum_1^5 \psi_i^2 + \frac{3}{5} \psi_6^2 + \frac{2}{5} \left[\psi_1 \psi_2 + \psi_3 (\psi_2 + \psi_4) + \psi_5 (\psi_4 + \psi_6) \right] \right\}.$$

Coefficient s_0 is calculated from eqn. (3.24). On integrating eqn. (3.33) by the trigonometric series method while retaining the first term, the free term of the equation corresponding to a given lateral loading is determined from eqn. (3.25). The following values are obtained for the basic data given in Section 3.1.5 of nodal displacements ψ_i using eqn. (3.14) for the shell under consideration at $\overline{P} = 1$:

$$\psi_1 = 1.100950 \times 10^{-4}, \quad \psi_2 = 2.443426 \times 10^{-4},$$

$$\psi_3 = 3.591758 \times 10^{-4}, \quad \psi_4 = 4.434005 \times 10^{-4},$$

$$\psi_5 = 4.961158 \times 10^{-4}, \quad \psi_6 = 5.172067 \times 10^{-4},$$

Then, $\quad a_0 = 759.0712 \times 10^{-8}, \quad s_0 = 0.0903125 \times 10^{-12},$

$$q_0(z) = 27.647597 \times 10^{-4} \frac{P}{G} \cdot \sin \frac{\pi z}{l}$$

After substituting $V(z) \times \sin \pi z / l$ in eqn. (3.33) and reducing to $\sin \pi z / l$, we derive a simple algebraic equation:

$$759.0712 \times 10^{-8} \frac{\pi^4}{60^4} V + 0.0903125 \times 10^{-12} V = 27 \cdot 647597 \times 10^{-4} \frac{P}{G} = 0,$$

hence $V = 0.46286072 \times 10^{-8} \dfrac{P}{G}.$

Let us calculate the displacements in the shell:

$$v(z,s) = V \cdot \sin \frac{\pi z}{l} \cdot \psi(s), \ u(z,s) = -\frac{\pi}{l} V \cos \frac{\pi z}{l} \cdot \varphi(s).$$

The transverse displacements will be maximum in the midsection at $z = l/2$. Let us describe rib displacements in this section:

$$v_i = V \cdot \psi_i.$$

Longitudinal displacements evidently attain maximum proportions in the extreme terminal sections of the shell at $z = 0$ and at $z = l$:

$$u(z,s) = \pm \frac{\pi}{l} V \cdot \varphi(s).$$

The longitudinal displacements of nodal points of these sections will be:

$$u_i = \pm \frac{\pi}{l} V \cdot \varphi_i.$$

The values at these characteristic points of the shell found with and without consideration of shear deformations are given in Tables 3.5 and 3.6.

Table 3.5 Transverse Displacements in the Midsection of Shell

Node	1	2	3	4	5	6
Allowing for shear	5600	13553	19923	24619	27519	28689
Ignoring shear	5096	11310	16624	20552	22963	23940
Deviation, %	9.7	16.5	16.5	16.5	16.5	16.5

Table 3.6 Longitudinal Displacements of Nodes in the Terminal (End) Sections

Node	1	2	3	4	5	6
Allowing for shear	893	1861	2718	3371	3810	4098
Ignoring shear	853	1777	2585	3195	3605	3769
Deviation, %	4.5	4.5	4.9	5.2	5.36	8.2

Note. The displacement values shown in the tables should be multiplied by P/G.

3.3.4 Determining the Range of Application of the Theory of Analysis Ignoring Shear Deformations

Determining accurate results in this respect is extremely difficult since the degree of influence of shear deformations on the stressed and deformed state of the class of structure under consideration depends on a host of factors, especially whether the strips forming the shell structure are continuous on the ratio of strip thickness and cell sizes (δ/d), on the size ratio of shells in the plan (l/b), on the ratio of the thickness and width of shell section (dz/b), on the ratio of cell dimensions (d_2/d_1) and on the type of material (reinforced concrete, sheet metal or corrugated metal composite with oriented fibres).

However, the investigations of Gol'denveizer (1949), Mileikovskii (1950a, 1950b, 1960), Vlasov and Mroshchinskii (1950), Vlasov (1958) and Meshcheryakov (1964, 1965) show that the main criteria in this respect are nevertheless the geometric parameters l/b and δ/d, in which l and b are the shell dimensions in the plan and δ and d the shell thickness and typical section size. Taking into consideration these recommendations, let us study the influence of shear deformations on displacements in shells formed of continuous flat members. Let us compare the extent of maximum deflection of the shell studied in the above examples (v_6 in the midsection) and maximum longitudinal nodal displacement (u_6 in the terminal section) at different ratios of $\delta/d = 1/75$, $1/100$ and $1/150$ and different ratios of shell dimensions in the plan $l/b = 1.67$, 2 and 3. For analysing, a shell of periodic structure allowing for shear deformations, eqn. (3.33) has been used. The results of calculations are shown graphically in the form of dependences of the ratios of maximum displacements in the shell calculated with and without allowance for shear deformations at different values of δ/d and $l/b = $ const. (Figs. 3.13 and 3.14) and also at different values of l/b and $\delta/d = $ const. (Figs. 3.15 and 3.16).

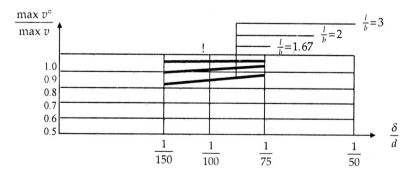

Fig. 3.13

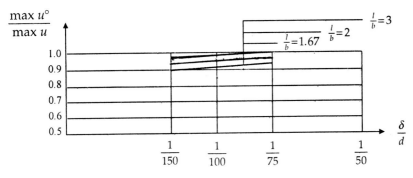

Fig. 3.14

The following conclusions emerge from the above analyses:

1) The main parameter determining the influence of shear deformation is the size ratio of shell l/b in the plan. Here, the following recommendation of Vlásov and Mroshchinskii (1950) and Mileikovskii (1950a, 1960) for cylindrical shells and folds have also been confirmed for prismatic shells with a multiconnected section: the impact of shear deformations increases as l/b decreases, especially noticeable in short shells at $l/b \leq 1$ to 1.5.

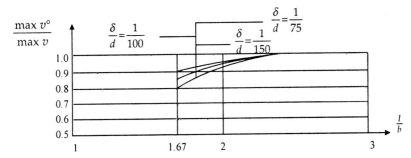

Fig. 3.15

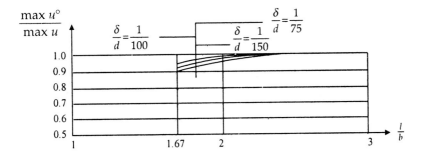

Fig. 3.16

2) In shells of medium length, the variation of parameter δ/d in a fairly wide range (1/150 to 1/50) had little influence on analytical accuracy when ignoring shear. It thus emerges that in shells of medium length comprising 'continuous' members, shear deformations may be ignored. They should be considered however in such cells in which the effective thickness of the shear zone is less than the effective thickness of the longitudinal tension zone of casings.

3.4 Theory of Analysis of Cellular Span Members with Multiconnected Section and Periodic Structure without Diaphragms under the Influence of Moving Concentrated Force

In bridge construction, there has recently been a transition from cellular span members of two or three cells to multiconnected, so-called wide or urban bridges. The available methods for analysing cellular spans are based on the theory of thin-walled closed sections. This has been dealt with in considerable detail in the works of Umanskii (1939), Luzhin (1959), Vlasov (1959), Biderman (1965), Il'yasevich (1970), Vol'nov (1978) and many others. The theory assumes the absence of deformations as the boundaries of the cellular section. This phenomenon necessitates saturating the span structure with a large number of transverse frames or diaphragms (Il'yasevich, 1970) since the assumption of contour non-deformability is relevant until the distance between the diaphragms does not exceed the section height (Birman, 1961). It is quite clear that this limitation calling for the construction of a large number of diaphragms makes the cellular construction needlessly expensive while at the same time depriving the structure of many useful qualities, such as the possibility of installing utility services and maintaining them at two levels.

These deficiencies become particularly manifest in the case of cellular spans with a multiconnected section. Evidently, the non-availability in the special 'bridge' literature of analytical methods free from hypotheses on contour non-deformability has partly restrained the adoption of multiconnected cellular spans without diaphragms. There is no doubt that ignoring diaphragms reduces the overall transverse stiffness of the structure. However, this stiffness reduction can be compensated when required by other structural measures, in particular by constructing composite type members.

3.4.1 *General Approach to the Problem*

A cellular span structure (Fig. 3.17) is a prismatic shell with a multiconnected section with two supporting and two free edges.

Restricting ourselves to the class of shells of medium length, let us apply Vlasov's variational method of displacements (1958), for which the differential equations are as follows:

$$\frac{E}{G}\sum a_{ji}\, U_i'' - \sum b_{ji}\, U_i - \sum c_{jk}\, V_k + \frac{1}{G}p_j = 0,$$

$$\sum c_{hi}\, U_i' + \sum \gamma_{hk}\, V_k'' - \frac{E}{G}\sum s_{hk}\, V_k + \frac{1}{G}g_h = 0, \qquad \ldots (3.35)$$

$$\left(i,\, j = 1,\, 2,\ldots,\, m = 2{\cdot}y;\ h,\, k = 1,\, 2,\ldots,\, n = 2{\cdot}y - c\right),$$

y and c being the number of nodes and shafts respectively in the hinged design of the shell cross-section.

The unknown longitudinal and transverse functions of displacement are represented in the following form:

$$u(z,s) = U_I\,(z){\cdot}\varphi_I(s) + U_{II}(z){\cdot}\varphi_{II}(s) + \sum_1^{m1} U_i(z){\cdot}\varphi_i(s),$$

$$v(z,s) = V_I\,(z){\cdot}\psi_I(s) + V_{II}(z){\cdot}\psi_{II}(s) + \sum_1^{n1} V_k(z){\cdot}\psi_k(s).$$

The following basic equations are obtained by opening up general eqn. (3.35) for the shell of multiconnected section under consideration:

$$EJ_x\, V_1^{IV}\,(z) - g_1(z) + \frac{EJ_x}{2GF_2}\, g_1(z) = 0;$$

$$a_k U_{II}''(z) - b_1 U_{II}(z) - b_2 V_{II}(z) = 0;$$

$$b_2 U_{II}'(z) + b_1 V_{II}''(z) + g_{II}(z) = 0;$$

$$\frac{E}{G}\sum a_{ji} U_i'' - \sum a_{ji} U_i - \sum c_{jk}\, V_k' = 0; \qquad \ldots (3.36)$$

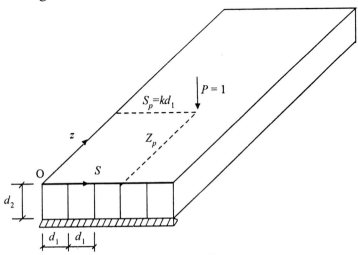

Fig. 3.17

84

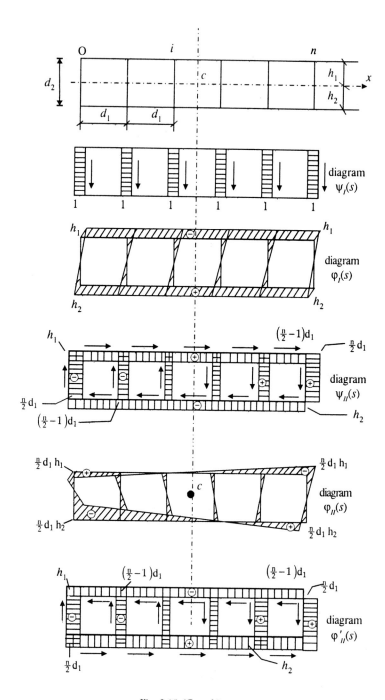

Fig. 3.18 (Contd.)

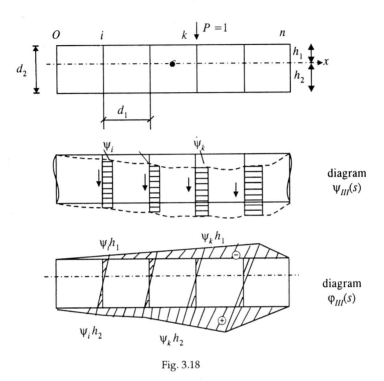

Fig. 3.18

$$\sum c_{hi} U_i' - \sum \gamma_{hk} V_k'' - \frac{E}{G} \sum s_{hk} V_k + \frac{1}{G} g_h = 0;$$

$$\left(j, i = 1, 2, \ldots, m_1 = m - 2\right), \left(h, k = 1, 2, \ldots, n_1 = n - 2\right)$$

The first equation, independent of the rest, defines the deformed state of the shell as it bends like a beam in the vertical plane. Under the action of concentrated loading, this equation coincides with that of the elementary theory of beam flexure. The next two equations of series (3.36) determine the deformed state of the shell which, for longitudinal displacements, is characterised by the warping of the section and, for transverse displacements, by the torsion angle of undeformed multiconnected contour. The subsequent equations form a system of order $(m_1 + n_1)$ and define the contour deformation which, in a multiconnected section, is characterised by additional longitudinal and transverse displacements of intermediate nodes of shell section.

Of the three special problems, bending, torsion and deformation of contour, the last is the most difficult. Its solution can be greatly simplified, however.

To resolve this special problem, let us use the method incorporated in Section 3.1. Ignoring the impact of shear deformations and introducing the single function of section contour deformation, the solution to the problem of span bending due to contour deformation can be reduced to a single differential equation:

$$a_0 V_{III}^{IV}(z) + s_0 V_{III}(z) = \frac{1}{E} q_{III}(z). \qquad \ldots (3.37)$$

Expressions of unknown displacements assume the following form:

$$u(z,s) = U_I(z).\varphi_I(s) + U_{II}(z).\varphi_{II}(s) + U_{III}(z).\varphi_{III}(s), \qquad \ldots (3.38)$$
$$v(z,s) = V_I(z).\psi_I(s) + V_{II}(z).\psi_{II}(s) + V_{III}(z).\psi_{III}(s).$$

Indices I, II and III denote the parameters pertaining respectively to the problems of bending, torsion and contour deformation.

Functions $\varphi_{III}(s)$ and $\psi_{III}(s)$, whose diagrams are depicted in Fig. 3.18, replace the families of functions $\varphi_i(s)$ and $\psi_k(s)$ (see Fig. 3.7); further, diagram $\varphi_{III}(s)$ is obtained on the basis of dependence (3.27) by 'integrating' $\psi_{III}(s)$ diagram: $\varphi_{III}(s) = \int \psi_{III}(s) \times ds$.

The unknown generalised longitudinal displacements $U(z)$ for given functions of $\varphi(s)$ are represented as: $U_I(z)$, the rotation angle of section $z = $ const. relative to axis OX; $U_{II}(z)$, the warping of section $z = $ const.; and $U_{III}(z)$, the additional longitudinal displacements of intermediate nodes of a multiconnected section under conditions of contour deformation. The unknown generalised transverse displacements $V(z)$ for given functions of $\psi(s)$ are:

$V_I(z)$, the vertical displacement (deflection) of section $z = $ const., $V_{II}(z)$, the rotation angle of section $z = $ const. as a rigid entity relative to longitudinal axis OZ; and $V_{III}(z)$, the additional lateral displacements of intermediate nodes of the multiconnected section under conditions of contour deformation.

The independent variables of equations are determined as generalised external forces at given displacements of an elementary strip of frame $\psi(s)$ in the following manner:

$$g_I(z) = P.\psi_I(s), \; g_{II}(z) = P.\psi_{II}(k), \; g_{III}(z) = P.\psi_{III}(k) \quad \text{at } z = z_p.$$

The coefficients of the above equations are:

$$a_k = E\int \varphi_{II}^2(s).dF, \qquad\qquad b_1 = G\int \psi_{II}^2(s).dF,$$

$$b_2 = G\int \psi_{II}(s).\varphi_{II}'(s).dF, \qquad a_0 = \int \varphi_{III}^2(s).dF,$$

$$S_0 = \int \int \left[\psi_{III}''(s)\right]^2 . ds .$$

3.4.2 Bending of a Span with Force P = 1

Ignoring here the influence of shear deformations on the extent of deflections as in the elementary theory of bending of a beam with a continuous wall and after integrating the first of eqns. (3.36), we find:

$$EJ_x V_{I(z)} = \begin{cases} -\dfrac{P}{6}\left[lz_p\left(1-\dfrac{z_p}{l}\right)\left(2-\dfrac{z_p}{l}\right)z - \left(1-\dfrac{z_p}{l}\right)z^3\right], & (0 \le z \le z_p); \\[3mm] \dfrac{P}{6}\left[\left(1-\dfrac{z_p}{l}\right)z^3 - (z-z_p)^3\right], & (z_p \le z \le l). \end{cases} \qquad \dots (3.39)$$

3.4.3 Torsion of a Span by Couple M = 1e

The type of torsion by a concentrated pair examined here presuming the absence of section contour deformation (Obraztsov, 1956 and Vlasov, 1958) is described by the second and third equations of system (3.36). As in Vlasov, 1958, let us introduce an auxiliary function f such that

$$U_{II}(z) = f', \quad V_{II}(z) = \frac{a_k}{b_2} f'' - \frac{b_1}{b_2} f.$$

Then the second equation is satisfied similarly and we obtain the following from the third equation after simple conversions:

$$f^{IV}(z) - \alpha^2 f''(z) = -\frac{b_2}{a_k b_1} q_{II}(z), \quad \text{here, } \alpha = \sqrt{\frac{b_1^2 - b_2^2}{a_k b_1}} \qquad \dots (3.40)$$

The general integral of a homogeneous equation is known:

$$f(z) = c_1 + c_2 \cdot z + c_3 e^{\alpha z} + c_4 e^{-\alpha z}. \qquad \dots (3.41)$$

Then, for the first section at $0 \le z_1 \le z_p$, we have

$$U_{II}(z_1) = c_2 + \alpha\left(c_3 e^{\alpha z_1} - c_4 e^{-\alpha z_1}\right),$$

$$V_{II}(z_1) = -\frac{b_1}{b_2}\left[c_1 + c_2 \cdot z_1 + \frac{b_2^2}{b_1^2}\left(c_3 e^{\alpha z_1} + c_4 e^{-\alpha z_1}\right)\right];$$

and for the second section at $z_p \le z_2 \le l - z_p$:

$$U_{II}(z_2) = D_2 + \alpha\left(D_3 e^{\alpha z_2} - D_4 e^{-\alpha z_2}\right),$$

$$V_{II}(z_2) = -\frac{b_1}{b_2}\left[D_1 + D_2 \cdot z_2 + \frac{b_2^2}{b_1^2}\left(D_3 e^{\alpha z_2} - D_4 e^{-\alpha z_2}\right)\right].$$

When the ends are hinge supported and the longitudinal margins are free, the limiting conditions will be:

$$z_1 = 0; \quad U_{II}'(z_1) = 0, \quad V_{II}(z_1) = 0,$$

$$z_1 = z_p; \quad U_{II}(z_1) = U_{II}(z_2), \ U'_{II}(z_1) = U'_{II}(z_2),$$

$$z_2 = 0; \ V_{II}(z_1) = U_{II}(z_2), \ b_2 \, U_{II}(z_1) + b_1 \, V_{II}{}'(z_1) - b_2 \, U_{II}(z_2) - b_1 \, V'_{II}(z_2) = M_k,$$

$$z_2 = l - z_p : \ U'_{II}(z_2) = 0, \ V_{II}(z_2) = 0.$$

Subjecting solution (3.41) to these conditions, we find:

$$C_1 = 0, \ C_2 = -A\left(1 - \frac{z_p}{l}\right), \ C_3 = -C_4 = \frac{A}{2\alpha} \frac{1 - e^{2\alpha(l - z_p)}}{F_1},$$

$$D_1 = -\frac{A}{\alpha}\left(1 - \frac{z_p}{l}\right) z_p,$$

$$D_2 = A\frac{z_p}{l}, \ D_3 = \frac{A}{\alpha} \cdot \frac{sh\,\alpha\,z_p}{F_1}, \ D_4 = -\frac{A}{\alpha} \cdot \frac{sh\,\alpha\,z_p}{F_1} l^{2\alpha(l - z_p)}$$

in which

$$A = \frac{b_2 \, M_k}{b_1^2 - b_2^2}, \ F_1 = \left[1 - e^{2\alpha(l - z_p)}\right] ch\,\alpha\,z_p - \left[1 + e^{2\alpha(l - z_p)}\right] sh\,\alpha\,z_p.$$

Then we finally derive:

$$U_{II}(z_1) = -\frac{b_2 \, M_k}{b_1^2 - b_2^2}\left[1 - \frac{z_p}{l} - \frac{1 - e^{2\alpha(l - z_p)}}{F_1} ch\,\alpha\,z_l\right],$$

$$V_{II}(z_1) = -\frac{b_1 \, M_k}{b_1^2 - b_2^2}\left[\left(1 - \frac{z_p}{l}\right) z_l - \frac{b_2^2}{\alpha b_1^2} \cdot \frac{1 - e^{2\alpha(l - z_p)}}{F_1} sh\,\alpha\,z_l\right],$$

$$U_{II}(z_2) = -\frac{b_2 \, M_k}{b_1^2 - b_2^2}\left[\frac{z_p}{l} + \frac{sh\,\alpha\,z_p}{F_1}\left(e^{\alpha z_2} + e^{2\alpha(l - z_p)} \cdot e^{-\alpha z_2}\right)\right], \quad \ldots \ (3.42)$$

$$V_{II}(z_2) = -\frac{b_1 \, M_k}{b_1^2 - b_2^2}\left[\left(1 - \frac{z_p}{l} - \frac{z_2}{l}\right) z_p - \frac{b_2^2}{\alpha b_1^2} \cdot \frac{sh\,\alpha\,z_p}{F_1} \times\right.$$

$$\left. \left(e^{\alpha z_2} + e^{2\alpha(l - z_p)} \cdot e^{-\alpha z_2}\right)\right],$$

in which $M_k = l \times e_k \times \psi_{II}(k)$ and k the number of the rib on which force has been applied.

3.4.4 Bending due to Section Contour Deformation under Force P = 1. Solution in Beam Fundamental Functions

The solution to this problem involves integrating eqn. (3.37) which coincides with the differential equation for beam bending on an elastic foundation. By dividing all its terms by a_0 and introducing the notation $\beta = \sqrt[4]{(S_0/4a_0)}$, we find:

$$V_{III}^{IV}(z) + 4\beta^4 V_{III}(z) = \frac{1}{Ea_0} q_{III}(z). \qquad \dots (3.43)$$

A general solution to the corresponding homogeneous equation is known:

$$V(z) = c_1 \sin \beta z \cdot sh\, \beta z + c_2 \sin \beta z \cdot ch\, \beta z + c_3 \cos \beta z \cdot sh\, \beta z + c_4 \cos \beta z \cdot ch\, \beta z.$$

To represent the solution in the form of the method of initial parameters, let us consider the following generalised factors at the beginning of the co-ordinates ($z = 0$): deflection $V(0)$, rotation angle $V'(0)$, bending moment $V''(0)$ and transverse force $V'''(0)$. Then, we find:

$$C_1 = \frac{V''(0)}{2\beta^2}, C_2 = \frac{V'(0)}{2\beta} + \frac{V'''(0)}{4\beta^3}, C_3 = \frac{V'(0)}{2\beta} - \frac{V'''(0)}{4\beta^3}, C_4 = V(0).$$

Later, by introducing the notation:

$$V'(0) = \varphi_0, Ea_0 V''(0) = M_0, Ea_0 V'''(0) = Q_0, V(0) = y_0,$$

the integration constants are expressed in terms of the initial parameters:

$$C_1 = \frac{1}{2}\overline{M}_0, C_2 = \frac{1}{2}\overline{\varphi}_0 + \frac{1}{4}\overline{Q}_0, C_3 = \frac{1}{2}\overline{\varphi}_0 - \frac{1}{4}\overline{Q}_0, C_4 = y_0,$$

in which

$$\overline{\varphi}_0 = \frac{\varphi_0}{\beta}, \overline{M}_0 = \frac{M_0}{Ea_0 \beta^2}, \overline{Q}_0 = \frac{Q_0}{Ea_0 \beta^3}. \qquad \dots (3.44)$$

Finally, using Krylov's notation (1931) for the beam fundamental functions:

$$A_{\beta z} = \cos \beta z \cdot ch\, \beta z, B_{\beta z} = \frac{1}{2}(\sin \beta z \cdot ch\, \beta z + \cos \beta z \cdot sh\, \beta z),$$

$$C_{\beta z} = \frac{1}{2}\sin Bz \cdot sh\, \beta z, D_{\beta z} = \frac{1}{4}(\sin \beta z \cdot ch\, \beta z - \cos \beta z \cdot sh\, \beta z),$$

we obtain for the generalised parameters in an arbitrary section of the shell:

$$V(z) = y_0 A_{\beta z} + \overline{\varphi}_0 B_{\beta z} + \overline{M}_0 C_{\beta z} + \overline{Q}_0 D_{\beta z},$$

$$V'(z) = \beta\left(-4y_0 D_{\beta z} + \overline{\varphi}_0 A_{\beta z} + \overline{M}_0 B_{\beta z} + \overline{Q}_0 C_{\beta z}\right), \qquad \dots (3.45)$$

$$V''(z) = \beta^2 \left(-4y_0 C_{\beta z} - 4\overline{\varphi}_0 D_{\beta z} + \overline{M}_0 A_{\beta z} + \overline{Q}_0 B_{\beta z} \right),$$

$$V'''(z) = \beta^3 \left(-4y_0 B_{\beta z} - 4\overline{\varphi}_0 C_{\beta z} - 4\overline{M}_0 D_{\beta z} + \overline{Q}_0 A_{\beta z} \right).$$

When force $P = 1$ is arbitrarily applied along the span length, the limiting conditions will be:

$$z_1 = 0: \quad V_{III}(z_1) = 0, \ V_{III}''(z_1) = 0,$$

$$z_1 = z_p: \quad V_{III}(z_1) = V_{III}(z_2), \ V_{III}'(z_1) = V_{III}'(z_2),$$

$$z_2 = 0: \quad V_{III}''(z_1) = V_{III}''(z_2), \ V_{III}'''(z_1) - V_{III}'''(z_2) = \frac{P \cdot \psi_{III}(k)}{Ea_0},$$

$$z_2 = l - z_p: \quad V_{III}(z_2) = 0, \ V_{III}''(z_2) = 0.$$

Satisfying these conditions, we find:

—for the first section $\left(0 \le z_1 \le z_p \right)$:

$$y_0^{(1)} = 0, \quad \overline{\varphi}_0^{(1)} = -\frac{P \cdot \psi_{III}(k)}{Ea_0 \beta^3} \cdot \frac{B_{\beta(l-z_p)} - \dfrac{b}{c} \cdot D_{\beta(l-z_p)}}{4c + b^2/c}, \qquad \dots (3.46)$$

$$\overline{M}_0^{(1)} = 0, \quad \overline{Q}_0^{(1)} = \frac{P \cdot \psi_{III}(k)}{Ea_0 \beta^3} \cdot \frac{1}{c} \times$$

$$\times \left[D_{\beta(l-z_p)} + \frac{b \cdot B_{\beta(l-z_p)} - \dfrac{b^2}{c} \cdot D_{\beta(l-z_p)}}{4c + b^2/c} \right], \qquad \dots (3.47)$$

—for the second section $\left(0 < z_2 \le l - z_p \right)$:

$$y_0^{(2)} = \overline{\varphi}_0^{(1)} B_{\beta z_p} + \overline{Q}_0^{(1)} \cdot D_{\beta z_p}, \quad \overline{\varphi}_0^{(2)} = \overline{\varphi}_0^{(1)} \cdot A_{\beta z_p} + \overline{Q}_0^{(1)} \cdot c_{\beta z_p},$$

$$\overline{M}_0^{(2)} = -4 \cdot \overline{\varphi}_0^{(1)} \cdot D_{\beta z_p} + \overline{Q}_0^{(1)} \cdot B_{\beta z_p},$$

$$\overline{Q}_0^{(2)} = -4 \cdot \overline{\varphi}_0^{(1)} \cdot c_{\beta z_p} + \overline{Q}_0^{(1)} \cdot A_{\beta z_p} - \frac{P \cdot \psi_{III}(k)}{Ea_0 \beta^3}, \qquad \dots (3.48)$$

in which

$$\overline{\varphi}_0 = \frac{\varphi_0}{\beta}, \quad \overline{M}_0 = \frac{M_0}{E \cdot a_0 \beta^2}, \quad \overline{Q}_0 = \frac{Q_0}{E \cdot a_0 \cdot \beta^3},$$

$$b = A_{\beta z_p} \cdot B_{\beta(l-z_p)} + B_{\beta z_p} \cdot A_{\beta(l-z_p)} - 4c_{\beta z_p} \cdot D_{\beta(l-z_p)} \times$$

$$- 4D_{\beta z_p} \cdot c_{\beta(l-z_p)}, \qquad \dots (3.49)$$

$$c = A_{\beta zp} \cdot D_{\beta(l-zp)} + D_{\beta zp} \cdot A_{\beta(l-zp)} + B_{\beta zp} \cdot c_{\beta(l-zp)} + C_{\beta zp} \cdot B_{\beta(l-zp)}.$$

The equations for the unknown generalised displacements in the first and second sections of the span structure assume the form:

$$V_{III}(z_1) = \overline{\varphi}_0^{(1)} \cdot B_{\beta z_1} + \overline{Q}_0^{(1)} \cdot D_{\beta z_1},$$

$$V_{III}(z_2) = y_0^{(2)} \cdot A_{\beta z_2} + \overline{\varphi}_0^{(2)} \cdot B_{\beta z_2} + \overline{M}_0^{(2)} \cdot c_{\beta z_2} + \overline{Q}_0^{(2)} \cdot D_{\beta z_2}. \qquad \text{... (3.50)}$$

3.4.5 Contour Deformation Function for a Multiconnected Section with a Periodic Structure

For this study, an elementary strip of frame is cut from the shell of the span structure such that its nodes corresponding to the longitudinal edges of the shell cannot undergo longitudinal displacements (Fig. 3.19).

Let us solve the problem of displacements. The equilibrium equation for arbitrary node i in the deformed state of the frame will be (Kornoukhov, 1949):

$$\sum M_i = \overline{M}_{i,i+1} + \overline{M}_{i,i-1} + \overline{M}_{i,i'} = 4\frac{EJ_1}{d_1}\,\overline{\omega}_i + 2\frac{EJ_1}{d_1}\,\overline{\omega}_{i+1} -$$

$$6\frac{EJ_1}{d_1}\,\overline{\Omega}_{i,i+1} + 4\frac{EJ_1}{d_1}\,\overline{\omega}_i + 2\frac{EJ_1}{d_1}\,\overline{\omega}_{i-1} - 6\frac{EJ_1}{d_1}\,\overline{\Omega}_{i,i-1} + 6\frac{EJ_3}{d_2}\,\overline{\omega}_{i'} = 0,$$

here, $\overline{\omega}_{i-1}, \overline{\omega}_i, \overline{\omega}_{i+1}, \overline{\omega}_{i'}$ are the rotation angles of nodes $i{-}1$, i, $i{+}1$ and i' respectively.

$\overline{\Omega}_{i,i-1}, \overline{\Omega}_{i,i+1}$ the rotation angles of the elements of the chord in panels i and $i{+}1$ i.e.,

$$\overline{\Omega}_{i,i-1} = \frac{1}{d_1}\left(\overline{\psi}_i - \overline{\psi}_{i-1}\right),\ \overline{\Omega}_{i,i+1} = \frac{1}{d_1}\left(\overline{\psi}_{i+1} - \overline{\psi}_i\right) \text{ and}$$

$\overline{\psi}_{i-1}, \overline{\psi}_i, \overline{\psi}_{i+1}$ are the vertical linear displacements of nodes $i{-}1$, i and $i{+}1$.

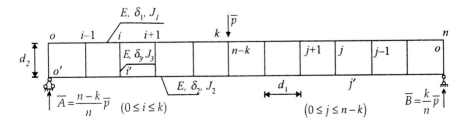

Fig. 3.19

Let us confine here to a cell wall of uniform thickness ($\delta_1 = \delta_2 = \delta$ and $J_1 = J_2 = J$). Then, the displacements of nodes in the lower chord of the frame will be inversely symmetrical to the displacements of the corresponding nodes on the upper chord, specially $\overline{\omega}_{i'} = \overline{\omega}_i$ and the equilibrium equation of the i-th node assumes a simple form:

$$2\frac{EJ}{d_1}\overline{\omega}_{i-1} + 2\frac{EJ}{d_1}\left(4 + 3\frac{J_3}{J}\frac{d_1}{d_2}\right)\overline{\omega}_i + 2\frac{EJ}{d_1}\overline{\omega}_{i+1} -$$

$$6\frac{EJ}{d_1}\left(\overline{\Omega}_{i,j-1} + \overline{\Omega}_{i,i+1}\right) = 0. \qquad \ldots (3.51)$$

The rotation angles of chords $\overline{\Omega}_{i,i-1}$ and $\overline{\Omega}_{i,i+1}$ can be excluded from eqns. (3.51). For this purpose, let us first examine the equilibrium of the left part of the frame up to node $i + 1$ (Fig. 3.20); it thus follows:

$$\overline{A} - 2\cdot\overline{Q}_{i,i+1} = 0, \text{ here, } \overline{Q}_{i,i+1} = -\frac{2EJ}{d_1^2}\left[3(\overline{\omega}_i + \overline{\omega}_{i+1}) - 6\overline{\Omega}_{i,i+1}\right].$$

After substituting $\overline{Q}_{i,i+1}$ and $\overline{A}$, we find:

$$\overline{\Omega}_{i,i+1} = \frac{1}{2},\left(\overline{\omega}_i + \overline{\omega}_{i+1}\right) + \frac{\overline{P}\,d_1^2}{24EJ}\cdot\frac{n-k}{n}. \qquad \ldots (3.52)$$

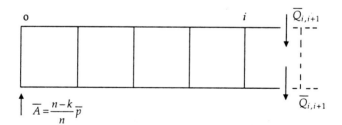

Fig. 3.20

Similarly, from the equilibrium of the left part of the frame up to node i, we have

$$\overline{\Omega}_{i,i-1} = \frac{1}{2}\left(\overline{\omega}_i + \overline{\omega}_{i-1}\right) + \frac{\overline{P}\,d_1^2}{24EJ}\cdot\frac{n-k}{n}. \qquad \ldots (3.53)$$

Now, the equilibrium equation for an arbitrary node assumes the following form:

$$-\overline{\omega}_{i-1} + 2\left(I + 3\frac{J_3}{J}\cdot\frac{d_3}{d_2}\right)\overline{\omega}_i - \overline{\omega}_{i+1} = \frac{\overline{P}\cdot d_1^2}{2EJ}\cdot\frac{n-k}{n}. \qquad \ldots (3.54)$$

For arbitrary node j situated to the right of point k of force $\overline{P}$ application, we can similarly derive:

$$-\overline{\omega}_{j-1} + 2\left(I + 3\frac{J_3}{J}\cdot\frac{d_1}{d_2}\right)\overline{\omega}_j - \overline{\omega}_{j+1} = -\frac{\overline{P}\cdot d_1^2}{2EJ}\cdot\frac{k}{n}. \qquad \text{... (3.55)}$$

Equations (3.54) and (3.55) represent inhomogeneous equations of finite differences of the second order. To solve them, use of the system fairly well developed by Bleikh and Melan (1938) and several other investigators in the field of structural mechanics (Rabinovich, 1921; Vol'vich, 1939; Kornoukhov, 1949; Segal', 1949; Ignatiev, 1973, 1979) yields a solution in a closed analytical form.

Following Bleikh and Melan (1938), we can derive a general solution to eqns. (3.54) and (3.55) in the form: $\overline{\omega}_i = \lambda^i$. Substitution in the left side of these equations gives the following characteristic equation:

$$\lambda^2 - 2\left(I + 3\frac{J_3}{J}\frac{d_1}{d_2}\right)\lambda + I = 0, \qquad \text{... (3.56)}$$

the roots of the above equation being

$$\lambda_1 = \frac{a}{2} + \sqrt{\frac{a^2}{4} - I} \quad \text{and} \quad \lambda_2 = \frac{a}{2} - \sqrt{\frac{a^2}{4} - I}.$$

Let us denote $\lambda_1 = \lambda$. Then, the roots assume a very simple form:

$$\lambda = \frac{a}{2} + \sqrt{\frac{a^2}{4} - I} \quad \text{and} \quad 1/\lambda, \text{ in which } a = 2\left(I + 3\frac{J_3}{J}\frac{d_1}{d_2}\right).$$

Conforming to the type of roots, the general solution will be:

$$\overline{\omega}_i^0 = C_1\,\lambda^i + C_2\,\lambda^{-i}, \quad \overline{\omega}_j^0 = D_1\,\lambda^j + D_2\,\lambda^{-j}.$$

The special solution of equilibrium equations can be found, however, in the following form:

$$\overline{\omega}_i^{00} = \left(A_1 + i\cdot B_1\right)\overline{P}, \quad \overline{\omega}_j^{00} = \left(A_2 + j\cdot B_2\right)\overline{P}.$$

We thus find:

$$\overline{\omega}_i^{00} = \frac{\overline{P}\,d_1\,d_2}{12\,EJ_3}\cdot\frac{n-k}{n}, \quad \overline{\omega}_j^{00} = \frac{\overline{P}\,d_1\,d_2}{12\,EJ_3}\cdot\frac{k}{n}.$$

Finally, we have:

$$\overline{\omega}_i = C_1\,\lambda^i + C_2\,\lambda^{-i} + \frac{\overline{P}\,d_1\,d_2}{12\,E\,J_3}\cdot\frac{n-k}{n},$$

$$\overline{\omega}_j = D_1\,\lambda^j + D_2\,\lambda^{-j} + \frac{\overline{P}\,d_1\,d_2}{12\,E\,J_3}\cdot\frac{k}{n}. \qquad\qquad \text{... (3.57)}$$

The equilibrium equation for the extreme left node, the continuity conditions of displacements in node k as also the equilibrium equation of the marginal right node on the frame can be used as the limiting conditions:

$$C_1\left(a - I - \lambda\right) + C_2\left(a - I - \frac{1}{\lambda}\right) = -\frac{\overline{P}\,d_1^2}{4EJ}\cdot\frac{n - k}{n},$$

$$C_1\,\lambda^k + C_2\lambda^{-k} - D_1\,\lambda^{n-k} - D_2\,\lambda^{k-n} = -\frac{\overline{P}\,d_1\,d_2}{12\,E\,J_3},$$

$$C_1\left(I + 2\frac{\lambda^k - \lambda}{\lambda - 1} + \lambda^k\right) + C_2\left(I + 2\frac{\lambda^k - \lambda}{\lambda - 1}\cdot\frac{1}{\lambda^k} + \frac{1}{\lambda^k}\right) +$$

$$D_1\left(I + 2\frac{\lambda^{n-k} - \lambda}{\lambda - 1} + \lambda^{n-k}\right) + D_2\left(I + 2\frac{\lambda^{n-k} - \lambda}{\lambda - 1}\,\lambda^{k-n} + \lambda^{k-n}\right) = 0, \quad \text{... (3.58)}$$

$$D_1\left(a - I - \lambda\right) + D_2\left(a - I - \frac{1}{\lambda}\right) = \frac{\overline{P}\,d_1^2}{4EJ}\cdot\frac{k}{n}.$$

Solving these equations, we find:

$$C_1 = -\frac{\overline{P}\,d_1\,d_2}{12\,E\,J_3}C, \quad C_2 = -\frac{\overline{P}\,d_1\,d_2}{12EJ_3}\left[\frac{1}{\lambda}\cdot C + \left(\frac{a}{2} - I\right)\frac{n - k}{n}\cdot\frac{1}{\lambda - 1}\right],$$

$$D_1 = -\frac{\overline{P}\,d_1\,d_2}{12\,E\,J_3}\cdot D, \quad D_2 = -\frac{\overline{P}\,d_1\,d_2}{12EJ_3}\left[\frac{1}{\lambda}\cdot D + \left(\frac{a}{2} - I\right)\frac{k}{n}\frac{1}{\lambda - 1}\right], \quad \text{... (3.59)}$$

here, $C = \dfrac{B}{\tilde{B}}$,

$$B = \frac{1}{\lambda^{n-k}}\Lambda_{(n-k)}\left(I + \frac{1}{\lambda}\cdot\frac{\Lambda_{k-n}}{\Lambda_{n-k}}\right)\left[I - \left(\frac{a}{2} - I\right)\frac{1}{\lambda - 1}\times\right.$$

$$\left(\frac{n - k}{n}\cdot\frac{1}{\lambda^k} + \frac{k}{n}\cdot\frac{1}{\lambda^{n-k}}\right)\right] - \left(\frac{a}{2} - I\right)\frac{1}{\lambda - 1}\times$$

$$\left(\frac{n - k}{n}\Lambda_{(-k)} - \frac{k}{n}\Lambda_{(k-n)}\right),$$

$$\tilde{B} = \Lambda_{(k)}\left(I + \frac{1}{\lambda}\cdot\frac{\Lambda_{(-k)}}{\Lambda_{(k)}}\right) + \frac{\lambda^{2k+1} + 1}{\lambda^{n+1}}\Lambda_{(n-k)}\left(I + \frac{1}{\lambda}\cdot\frac{\Lambda_{(k-n)}}{\Lambda_{(n-k)}}\right) \qquad \text{... (3.60)}$$

$$D = \frac{1}{\lambda^{n-k}}\left[-C\cdot\lambda^k + I - \left(\frac{a}{2} - I\right)\left(\frac{n-k}{n}\cdot\frac{1}{\lambda^k} + \frac{k}{n}\cdot\frac{1}{\lambda^{n-k}}\right)\frac{1}{\lambda-1}\right],$$

$$\Lambda_{(-k)} = I + 2\frac{\lambda^k - \lambda}{\lambda-1}\cdot\frac{1}{\lambda^k} + \frac{1}{\lambda^k}, \qquad \Lambda_{(k)} = I + 2\frac{\lambda^k - \lambda}{\lambda-1} + \lambda^k,$$

$$\Lambda_{(k-n)} = I + 2\frac{\lambda^{n-k} - \lambda}{\lambda-1}\cdot\frac{1}{\lambda^{n-k}} + \frac{1}{\lambda^{n-k}}, \qquad \Lambda_{(n-k)} = I + 2\frac{\lambda^{n-k} - \lambda}{\lambda-1} + \lambda^{n-k}.$$

Then, the angular and linear displacements of nodes on the frame will be:

$$\overline{\omega}_i = \frac{\overline{P}\,d_1\,d_2}{12EJ_3}\left\{\frac{n-k}{n} - C\lambda^i - \left[\frac{1}{2}C + \left(\frac{a}{2} - I\right)\frac{n-k}{n}\cdot\frac{l}{\lambda-l}\right]\frac{1}{\lambda^i}\right\},$$

$$\overline{\omega}_j = \frac{\overline{P}\,d_1\,d_2}{12EJ_3}\left\{\frac{k}{n} - D\lambda^j - \left[\frac{1}{\lambda}D + \left(\frac{a}{2} - I\right)\frac{k}{n}\cdot\frac{l}{\lambda-l}\right]\frac{1}{\lambda^j}\right\}. \quad \dots\ (3.61)$$

$$\overline{\psi}_i = \frac{\overline{P}\,d_1^2\,d_2}{12EJ_3}\left\{\frac{a+10}{6}\cdot\frac{n-k}{n}\cdot i - C\left(I + 2\frac{\lambda^i - \lambda}{\lambda-1} + \lambda^i\right) - \right.$$

$$\left. \left[\frac{1}{\lambda}C + \left(\frac{a}{2} - I\right)\frac{n-k}{n}\cdot\frac{1}{\lambda-1}\right]\left(I + 2\frac{\lambda^i - \lambda}{\lambda-1}\cdot\frac{1}{\lambda^i} + \frac{1}{\lambda^i}\right). \quad \dots\ (3.62)\right.$$

$$\overline{\psi}_j = \frac{\overline{P}\,d_1^2\,d_2}{24\,EJ_3}\left\{\frac{a+10}{6}\cdot\frac{k}{n}\cdot j - D\left(I + 2\frac{\lambda^j - \lambda}{\lambda-1} + \lambda^j\right) - \right.$$

$$\left. \left[\frac{1}{\lambda}D + \left(\frac{a}{2} - 1\right)\cdot\frac{k}{n}\cdot\frac{1}{\lambda-1}\right]\left(I + 2\frac{\lambda^j - \lambda}{\lambda-1}\cdot\frac{1}{\lambda^j} + \frac{1}{\lambda^j}\right).\right.$$

Dimensionless expressions contained in the braces of equation (3.62) can conveniently be used as an approximating function of contour deformation:

$$\psi_{III}(i) = \frac{a+10}{6}\cdot\frac{n-k}{n}\cdot i - C\left(I + \frac{\lambda^i - \lambda}{\lambda-1} + \lambda^i\right) - \quad \dots\ (3.63)$$

$$(0 \le i \le k) \qquad \left[\frac{1}{\lambda}C + \left(\frac{a}{2} - 1\right)\frac{n-k}{n}\cdot\frac{1}{\lambda-1}\right]\left(I + 2\frac{\lambda^i - \lambda}{\lambda-1}\cdot\frac{1}{\lambda^i} + \frac{1}{\lambda^i}\right).$$

$$\overline{\Psi}_{III}(j)= \left\{ \frac{a+10}{6}\frac{k}{n}\cdot j - D\left(I+2\frac{\lambda^j-\lambda}{\lambda-1}+\lambda^j\right)-\right.$$

$$\left[\frac{1}{\lambda}D+\left(\frac{a}{2}-I\right)\cdot\frac{k}{n}\cdot\frac{1}{\lambda-1}\right]\left(I+2\frac{\lambda^j-\lambda}{\lambda-1}\cdot\frac{1}{\lambda^j}+\frac{1}{\lambda^j}\right).$$

3.4.6 Formulating Coefficient S_0 of Differential Equation in the Problem of Deformation of Contour with a Multiconnected Section

Integration in equation $S_0 = \int\int[\psi''_{III}(s)]^2 ds$ should be carried out for all the elements of the cross-section. For a shell with casings of identical thickness:

$$S_0 =2\sum_{i=1}^{k} \int\int_0^{d_1}[\psi''_i(s)]^2\cdot ds + 2\sum_{j=1}^{n-k}\int\int_0^{d_1}[\psi''_j(s)]^2\cdot ds + \sum_{i=0}^{k}J_3\int_0^{d_2}[\tilde{\psi}''_i(s)]^2\cdot ds +$$

$$\sum_{j=0}^{n-k-1}J_3\int_0^{d_2}[\tilde{\psi}''_j(s)]^2\cdot ds. \qquad\qquad \text{... (3.64)}$$

Index III of function $\psi(s)$ in the above equation has been deleted for the time being to simplify the expression. Function $\psi''_i(s)$ represents the dimensionless part of the second derivative from the equation for the bent axis of the i-th chord of the panel of the elementary strip of the frame (Fig. 3.21).

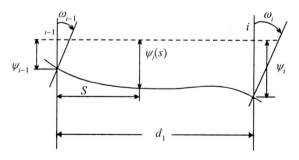

Fig. 3.21

$$\psi''_i(s) = 2\cdot\omega_{i-1} + \omega_i - \frac{3}{d_1}\left(\psi_i - \psi_{i-1}\right) -$$

$$\frac{6}{d_1}\left[\omega_{i-1}+\omega_i - \frac{2}{d_1}(\psi_i - \psi_{i-1})\right]\cdot s. \qquad\qquad \text{... (3.65)}$$

After substituting here the dimensionless expressions for linear and angular displacements of nodes on the frame, we find

$$\left[\psi_i''(s)\right]^2 = \frac{1}{d_1^2}\left[A_i^2 + 2A_i \cdot B_i \cdot s + B_i^2 \cdot s^2\right], \qquad \ldots (3.66)$$

in which

$$A_i = -\left(\frac{a}{2}-I\right)\frac{n-k}{n} + C(\lambda - I)\lambda^{i-1} - \left[\frac{1}{\lambda}C + \left(\frac{a}{2}-I\right)\frac{n-k}{n}\cdot\frac{1}{\lambda-1}\right]\times$$

$$\frac{\lambda - 1}{\lambda^i} \qquad \ldots (3.67)$$

$$B_i = -\frac{1}{d_1}(a-2)\cdot\frac{n-k}{n}.$$

Similarly, for panels located to the right of the loaded node k, we have:

$$\left[\psi_j''(s)\right]^2 = \left[A_j^2 + 2A_j \cdot B_j \cdot s + B_j^2 \cdot s^2\right]\frac{1}{d_1^2}, \qquad \ldots (3.68)$$

in which

$$A_j = \left(\frac{a}{2}-I\right)\frac{k}{n} - D(\lambda-I)\lambda^{j-1} + \left[\frac{1}{\lambda}D + \left(\frac{a}{2}-1\right)\frac{k}{n}\cdot\frac{1}{\lambda-1}\right]\times$$

$$\frac{\lambda - 1}{\lambda^j},$$

$$B_j = -\frac{1}{d_1}(a-2)\cdot\frac{k}{n}. \qquad \ldots (3.69)$$

Function $\tilde{\psi}_i''(s)$ for an arbitrary column (Fig. 3.22) has simpler form:

$$\psi_i'(s) = \tilde{A}_i - \frac{2}{d_2}\cdot\tilde{A}_i \cdot s.$$

Then, $\left[\tilde{\psi}_i''(s)\right]^2 = \tilde{A}_i^2\left(I - \frac{4}{d_2}\cdot s + \frac{4}{d_2^2}\cdot s^2\right), \qquad \ldots (3.70)$

here,

$$\tilde{A}_i = 6\frac{d_1}{d_2}\left\{\frac{n-k}{n} - C\lambda^i - \left[\frac{1}{\lambda} - C + \left(\frac{a}{2}-I\right)\frac{n-k}{n}\cdot\frac{1}{\lambda-1}\right]\frac{1}{\lambda^i}\right\}. \qquad \ldots (3.71)$$

For columns to the right of loaded node k:

$$\left[\tilde{\psi}_j''(s)\right]^2 = \tilde{A}_j^2\left(I - \frac{4}{d_2}\cdot s + \frac{4}{d_2^2}\cdot s\right), \qquad \ldots (3.72)$$

here, $\tilde{A}_j = 6\frac{d_1}{d_2}\left\{\frac{k}{n} - D\lambda^j - \left[\frac{1}{\lambda}D + \left(\frac{a}{2}-I\right)\frac{k}{n}\cdot\frac{1}{\lambda-1}\right]\frac{1}{\lambda^j}\right\}. \qquad \ldots (3.73)$

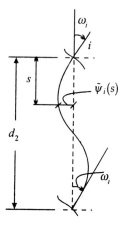

Fig. 3.22

After substitution and calculation of certain integrals, we derive:

$$S_0 = \frac{8J}{d_1^3}\left[\sum_{i=1}^{k}\left(A_i^2 + A_i \cdot B_i d_1 + \frac{d_1^2}{3} B_i^2\right) + \sum_{j=1}^{n-k}\left(A_j^2 + A_j \cdot B_j d_1 + \frac{d_1^2}{3} B_i^2\right) + \right.$$

$$\left. \frac{1}{3} \cdot \frac{J_3}{J} \cdot \frac{d_1}{d_2}\left(\sum_{i=1}^{k}\tilde{A}_i^2 + \sum_{j=0}^{n-k-1}\tilde{A}_j^2\right)\right],$$

$\qquad$ ... (3.74)

here, $\quad J = \frac{1 \cdot \delta^3}{12}, \, J = \frac{1 \cdot \delta_3^3}{12}, \, \frac{J_3}{J} = \left(\frac{\delta_3}{\delta}\right)^3.$

$\qquad$ ... (3.75)

3.4.7 Vertical Displacements (Deflections)

The total deflection of an arbitrary point on a span structure is determined by the overall influence of bending, torsion and deformation of the section contour:

$$v(z,s) = V_I(z)\cdot\psi_I(s) + V_{II}(z)\cdot\psi_{II}(s) + V_{III}(z)\cdot\psi_{III}(s).$$

Thus for deflections of points in the first section $\left(0 < z_l \leq z_p\right)$, we find:

$$v(z_1,s) = \frac{P}{6EJ_x}\left\{\left[-\left(2 - \frac{z_p}{l}\right)l\cdot z_p + z_1^2\right]\left(1 - \frac{z_p}{l}\right)z_1\right\}\psi_I(s) + \frac{b_1\,M_k}{b_1^2 - b_2^2}\times$$

$$\left[\left(1 - \frac{z_p}{l}\right)z_1 - \frac{b_2^2}{\alpha\,b_1^2}\cdot\frac{1 - e^{2\alpha(l-z_p)}}{F_1}\cdot sh\,\alpha z_1\right]\times$$

$$\psi_{II}(s) + \left(\overline{\varphi}_0^{(1)} \cdot B_{\beta z_1} + \overline{Q}_0^{(1)} \cdot D_{\beta z_1}\right) \cdot \psi_{III}(s). \qquad \text{... (3.76)}$$

and for deflection in the second section $\left(0 < z_2 \leq l - z_p\right)$:

$$v(z_2, s) = \frac{P}{6EJ_x}\left[-\left(1 - \frac{z_p}{l}\right)\left(2 - \frac{z_p}{l}\right) \cdot l \cdot z_p \cdot (z_p + z_2) + \left(1 - \frac{z_p}{l}\right) \times \right.$$

$$\left. (z_p + z_2)^3 - z_2^3\right]\psi_I(s) - \qquad \text{... (3.77)}$$

$$\frac{b_1 M_k}{b_1^2 - b_2^2}\left[\left(1 - \frac{z_p}{l} - \frac{z_2}{l}\right)z_p - \frac{b_2^2}{\alpha b_1^2} \cdot \frac{\text{sh } \alpha z_p}{F_1}\left(e^{\alpha z_2} - e^{2\alpha(l-z_p)} \cdot e^{-\alpha z_2}\right)\right] \times$$

$$\psi_{II}(s) + \left(y_0^{(2)} \cdot A_{\beta z_2} + \overline{\varphi}_0^{(2)} \cdot B_{\beta z_2} + \overline{M}_0^{(2)} \cdot C_{\beta z_2} + \overline{Q}_0^{(2)} \cdot D_{\beta z_2}\right)\psi_{III}(s).$$

3.4.8 Longitudinal Normal Stresses

Within the framework of V.Z. Vlasov's theory of shells of medium length, longitudinal normal stresses at points on the midsurfaces forming the shell of strips are determined as:

$$\sigma(z, s) = E \sum U'(z) \cdot \varphi(s).$$

In problems on bending and deformation of the section contour of a cellular span structure ignoring shear effect, $\psi(s) = \varphi'(s)$ and, hence $U(z) = -V'(z)$. Differentiating the latter for z, we have:

$$U'(z) = -V''(z). \qquad \text{... (3.78)}$$

In the torsion problem, the nature of a fixed support is such that the generalised transverse force $Q_{II}(z) = b_1 U_{II}(z) + b_2 V'_{II}(z)$ at all intermediate sections is not zero. This demonstrates that at the boundary conditions caused by torsion, bimoment normal stresses arise in transverse sections of the span. Therefore, allowing for eqn. (3.42), the equation for normal stresses assumes the form:

$$\sigma(z,s) = E\left[-V_I''(z) \cdot \varphi_I(s) + U_{II}'(z) \cdot \varphi_{II}(s) - V_{III}''(z) \cdot \varphi_{III}(s)\right]. \qquad \text{... (3.79)}$$

After determining the slopes $\dfrac{\partial v}{\partial z}$ and $\dfrac{\partial v}{\partial z}$ from eqns. (3.77), (3.76) and (3.39) and introducing the results in eqn. (3.79), we find eqn. (3.80) in which J_x is the moment of inertia of the transverse section relative to the horizontal central axis.

$$\sigma(z_1, s) = -\frac{P}{J_x}\left(1 - \frac{z_p}{l}\right) \cdot z_1 \cdot \varphi_I(s) + \frac{\alpha b_2 M_k}{b_1^2 - b_2^2} \cdot \frac{1 - e^{2\alpha(l - z_p)}}{F_1} \cdot \text{sh } \alpha z \cdot \varphi_{II}(s) -$$

$$E\beta^2\left(-4\overline{\varphi}_0^{(1)} D_{\beta z_1} + \overline{Q}_0^{(1)} B_{\beta z_1}\right)\varphi_{III}(s), \qquad \text{... (3.80)}$$

$$\sigma(z_2,s) = -\frac{P}{J_x}\left[\left(1-\frac{z_p}{l}\right)(z_p+z_2)-z_2\right]\varphi_I(s) + \frac{\alpha\, b_2\ M_k}{b_1^2-b_2^2}\cdot\frac{sh\,\alpha\,z_p}{F_1}\times$$

$$\left(e^{\alpha z_2} - e^{2\alpha(l-z_p)}\cdot e^{-\alpha z_2}\right).$$

$$\varphi_{II}(s) - E\beta^2\left(-4y_0^{(2)}\cdot C_{\beta z_2} - 4\overline{\varphi}_0^{(2)}\,D_{\beta z_2} + \overline{M}_0^{(2)}\,A_{\beta z_2} + \overline{Q}_0^{(2)}\,B_{\beta z_2}\right).$$

3.4.9 *Shear (Tangential) Stresses*

Since the problems of flexure and contour deformation of span type structures have been solved here ignoring shear deformations, shear stresses should be determined as in the elementary theory of beam bending, not on the basis of Hooke's law but from static considerations. Using V.Z. Vlasov's conclusions (1958), let us draw the equation for shear stresses in a strip falling between ribs $(i-1)$ and i:

$$\tau(z,s) = \frac{d_1}{6}\left[\sigma'_{i-1}(z)\cdot\left(2-\frac{6}{d_1}s+3\frac{s^2}{d_1^2}\right) + \sigma'_i(z)\cdot\left(1-\frac{3}{d_1^2}\cdot s^2\right)\right],$$

$$(0 \le s \le d_1). \qquad\qquad \dots (3.81)$$

During torsion of the span structure, shear stresses are found from the Hooke's law as:

$$\tau(z,s) = G\left(\frac{\partial v}{\partial z}+\frac{\partial u}{\partial s}\right) = G\left[U_{II}(z)\cdot\varphi'_{II}(s)+V'_{II}(z)\cdot\psi_{II}(s)\right]. \qquad \dots (3.82)$$

The first term in eqn. (3.46) takes into account the warping of transverse sections and the second their relative torsion.

Combining the effect of bending, torsion and contour deformation, we derive an equation for shear stresses in the nodal points of the lateral section of the span structure:

$$\tau(z_1,i) = -\frac{P}{2J_x}\left(1-\frac{z_p}{l}\right)\varphi_I(i) + G\frac{b_2\,M_k}{b_1^2-b_2^2}\times$$

$$\left\{\left(1-\frac{z_p}{l}\right)\left[-\varphi'_{II}(i)+\frac{b_1}{b_2}\,\psi_{II}(i)\right] - \frac{1-e^{2\alpha(l-z_p)}}{F_1}\cdot ch\,\alpha z_1\left[-\varphi'_{II}(i)+\frac{b_2}{b_1}\,\psi_{II}(i)\right]\right\}+$$

$$E\,\frac{d_1}{6}\,\beta^3\left(-4\overline{\varphi}_0^{(1)}C_{\beta z_1}+\overline{Q}_0^{(1)}\,A_{\beta z_1}\right)\left[\varphi_{III}(i-1)+2\,\varphi_{III}(i)\right],$$

$$\tau(z_2, i) = \frac{P}{2J_x} \cdot \frac{z_p}{l} \, \varphi_I(i) + G \frac{b_2}{b_1^2 - b_2^2} \frac{M_k}{} \left\{ -\frac{z_p}{l} \left[-\varphi'_{II}(i) + \frac{b_1}{b_2} \, \psi_{II}(i) \right] - \right.$$

$$\frac{sh\,\alpha\,z_p}{F_1} \left(e^{\alpha z_2} + e^{2\alpha(l-z_p)} \cdot e^{-\alpha z_2} \right) \times$$

$$\left[-\varphi'_{II}(i) + \frac{b_1}{b_2} \, \psi_{II}(i) \right] \right\} + E \frac{d_1}{6} \beta^3 - 4 y_0^{(2)} B_{\beta z_2} - 4 \overline{\varphi}_0^{(2)} C_{\beta z_2} -$$

$$4 \overline{M}_0^{(2)} D_{\beta z_2} + \overline{Q}_0^{(2)} A_{\beta z_2} \Big] \Big[\varphi_{III}(i-1) + 2\varphi_{III}(i) \Big]. \qquad \dots (3.83)$$

3.4.10 Transverse Deflection Moments

The transverse deflection moments in the elements of a cellular section arising as a result of its contour deformation can be found from the equation

$$M(z,s) = -V_{III}(z) \cdot M(s). \qquad \dots (3.84)$$

Equations for $M(s)$ are obtained within the framework of auxiliary problems studied in Section 3.3.5. Thus, the nodal deflection moment in strip of frame is:

$$M_{i,i+1} = \frac{4EJ}{d_1} \overline{\omega}_i + \frac{2EJ}{d_1} \overline{\omega}_{i+1} - \frac{6EJ}{d_1} \overline{\Omega}_{i,i+1} = \frac{4EJ}{d_1} \overline{\omega}_i + \frac{2EJ}{d_1} \overline{\omega}_{i+1} - \frac{6EJ}{d_1^2} \times$$

$$(\overline{\Psi}_{i+1} - \overline{\Psi}_i) = \frac{2EJ}{d_1} \left[2\overline{\omega}_i + \overline{\omega}_{i+1} - \frac{3}{d_1} (\overline{\Psi}_{i+1} - \overline{\Psi}_i) \right].$$

Substituting here the dimensionless part of the equations for the rotation angles of nodes from eqn. (3.61), we derive the function for the distribution of nodal moments:

$$M_{i,i+1} = \frac{2EJ}{d_1} \left\{ 3\frac{n-k}{n} - C\lambda^i(2+\lambda) - \frac{1}{\lambda^i} \left[\frac{1}{\lambda}C + \left(\frac{a}{2} - I \right) \frac{n-k}{n} \cdot \frac{1}{\lambda-1} \right] \times \right.$$

$$\left. \left(2 + \frac{1}{\lambda} \right) - \frac{3}{2}(\psi_{i+1} - \psi_i) \right\}. \qquad \dots (3.85)$$

The distribution function of nodal moments in sections to the left of nodes will be:

$$M_{i,i-1} = \frac{2EJ}{d_1} \left\{ 3\frac{n-k}{n} - C\lambda^i \left(2 + \frac{1}{\lambda} \right) - \frac{1}{\lambda^i} \left[\frac{1}{\lambda}C + \left(\frac{a}{2} - I \right) \frac{n-k}{n} \cdot \frac{1}{\lambda-1} \right] \times \right.$$

$$\left. (2 + \lambda) - \frac{3}{2} (\psi_i - \psi_{i-1}) \right\}. \qquad \dots (3.86)$$

Similarly, in the nodes falling to the right of the loaded node k will be:

$$M_{j,j+1} = \frac{2EJ}{d_1} \left\{ 3\frac{k}{n} - D\lambda^j (2+\lambda) - \frac{1}{\lambda^j} \left[\frac{1}{\lambda}D + \left(\frac{a}{2} - I\right)\frac{k}{n} \cdot \frac{1}{\lambda-1} \right] \times \right.$$

$$\left. \left(2 + \frac{1}{\lambda}\right) - \frac{3}{2} \left(\psi_{j+1} - \psi_j\right) \right\},$$

$$M_{j,j-1} = \frac{2EJ}{d_1} \left\{ 3\frac{k}{n} - D\lambda^j \left(2+\frac{1}{\lambda}\right) - \frac{1}{\lambda^j} \left[\frac{1}{\lambda}D + \left(\frac{a}{2} - I\right)\frac{k}{n} \cdot \frac{1}{\lambda-1} \right] \times \right.$$

$$\left. (2 + \lambda) - \frac{3}{2} \left(\psi_j - \psi_{j-1}\right) \right\}. \qquad \text{... (3.87)}$$

When deflection moments are required to be determined at points on the cell wall between nodes, the distribution function can be derived from Fig. 3.21:

$$M_{i,i-1}(s) = \frac{2EJ}{d_1} \left[2\omega_{i-1} + \omega_i - 3(\psi_i - \psi_{i-1}) - \frac{6EJ}{d_1^2} \times \right. \qquad \text{... (3.88)}$$

$$\left. \left[\omega_{i-1} + \omega_i - \frac{2}{d_1}(\psi_i - \psi_{i-1}) \right] \cdot s.$$

Thus, by substituting in eqn. (3.84) $V_{III}(z)$ from eqn. (3.50), we derive the final equations for transverse deflection moments in the first and second sections of the type structure:

$$M(z_1,s) = -\left(\overline{\varphi}_0^{(1)} \cdot B_{\beta z_1} + \overline{Q}_0^{(1)} \cdot D_{\beta z_1}\right) \cdot M(s), \qquad \text{... (3.89)}$$

$$M(z_2,s) = -\left(\overline{y}_0^{(2)} \cdot A_{\beta z_2} + \overline{\varphi}_0^{(2)} \; B_{\beta z_2} + \overline{M}_0^{(2)} \cdot C_{\beta z_2} + \overline{Q}_0^{(2)} \cdot D_{\beta z_2}\right) \cdot M(s).$$

Here, $M(s)$ represents the distribution function of deflection moments determined using eqns. (3.85) to (3.88).

3.5 Application of Theory for Constructing Influence Surfaces in a Cellular Span Structure

3.5.1 Basic Data

As an example of application of the theory, let us construct the influence surfaces for an urban single-span metal bridge without diaphragms. The length of the bridge $l = 75$ m and its cross-section is formed of a continuous closed shell in five sections with a total width $b = 17.5$ m (Fig. 3.23). A prototype of this span structure design is the cellular bridge in three sections described by Prof. S.A. Il'yasevich (1970) in the conventional manner, i.e. ignoring the contour deformation. Apart from increasing the number of sections in the cross-section, the present example has adopted identical thicknesses of upper and lower sheets of the compartment and its walls ($\delta_b = \delta_h = \delta_c = \delta = 0.014$ m). All other characteristic dimensions are maintained as before. It has been assumed that the outer walls are 0.4 m thicker than the central ones and the cantilevers have been fixed to the outer walls of the compartment. The height of intermediate walls is 3 m and end ones 3.4 m.

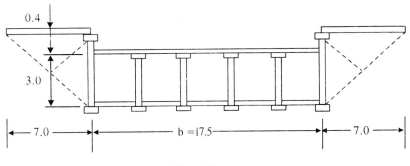

Fig. 3.23

The adopted analytical scheme of the section (Fig. 3.24) was obtained by concentrating the areas ΔF of the boom sheets of walls and also the cantilever sheets at points corresponding to the top and bottom of the corresponding walls.

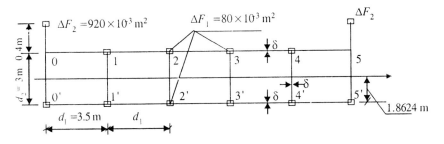

Fig. 3.24

The cross-section $F = 1.0172$ m² and moment of inertia relative to the horizontal central axis $J_x = 2.052063$ m⁴.

3.5.2 $\psi(s)$ and $\varphi(s)$ Diagrams in Bending and Torsion Problems

Function $\psi_j(s)$ describing the displacement of an elementary strip of frame in the plane of a cross-section under conditions of bending of the span structure is shown in Fig. 3.23, which also shows the diagram of displacements of the strip from the plane of cross-section $\varphi_j(s)$.

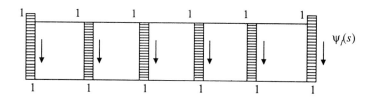

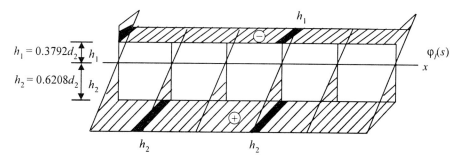

Fig. 3.25

Function $\psi_{II}(s)$ of displacements of the elementary frame in the plane of the section for solving the torsion problem is derived as the result of rotation of the section on a common angle (Fig. 3.26). Displacements of the strip of the frame from the plane of the section are fixed by function $\varphi_{II}(s)$ whose diagram is also shown in Fig. 3.26, as is the diagram of the first derivative of function $\varphi_{II}(s)$ along co-ordinates.

3.5.3 Constructing $\psi(s)$ and $\varphi(s)$ Diagrams in the Problem of Contour Deformation for a Multiconnected Section

The ordinates of diagram $\psi_{III}(s)$ represent the dimensionless nodal displacements in the strip of the frame and are determined using eqns. (3.63). Let us calculate the constant therein:

$$\frac{J_3}{J} = \frac{l \cdot \delta_3^3}{12} \cdot \frac{12}{l \cdot \delta_3^3} = 1, \quad a = 2\left(1 + 3\frac{J_3}{J} \cdot \frac{d_1}{d_2}\right) = 2\left(1 + 3 \cdot l \cdot \frac{3 \cdot 5}{5}\right) = 9,$$

$$\lambda = \frac{a}{2} + \sqrt{\frac{a^2}{4} - l} = 8.8875; \quad \frac{l}{\lambda} = 0.1125.$$

In the case under consideration, $n = 5$.

Assuming that $k = 1$, which corresponds to the loading of node No. 1 of the strip of the frame, we find from eqns. (3.59) and (3.60) that

$$\Lambda_{(k)} = 9.8875, \quad \Lambda_{(-k)} = 1.1125, \quad \Lambda_{(k-n)} = 7818, \quad \Lambda_{(k-n)} = 1.25296,$$

$$B = 0.918508, \quad P = 21.157577, \quad C = 0.04341, \quad D = 0.6666918716,$$

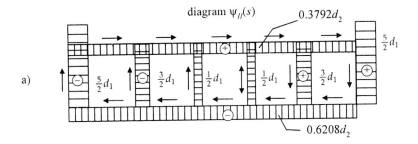

diagram $\psi_{II}(s)$

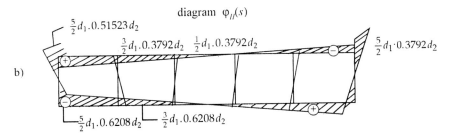

diagram $\varphi_{II}(s)$

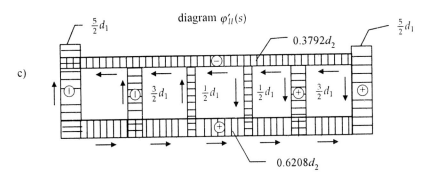

diagram $\varphi'_{II}(s)$

Fig. 3.26

and later from eqn. (3.63), we derive

$$\psi_1 = \psi_{(i=1)} = \psi_{(j=4)} = 1.706016, \quad \psi_2 = \psi_{(j=3)} = 1.147947,$$

$$\psi_3 = \psi_{(j=2)} = 1.7081847, \quad \psi_4 = \psi_{(j=1)} = 0.5336903.$$

At $k = 2$, we have from eqns. (3.24) and (3.27)

$$\Lambda_{(k)} = 97.7626, \quad \Lambda_{(-k)} = 1.2376084, \quad \Lambda_{(n-k)} = 878.74, \quad \Lambda_{(k-n)} = 1.216477,$$

$$B = 0.08801929, \quad P = 196.7753, \quad C = 0.0045188, \quad D = 0.0009108,$$

$$\psi_1 = \psi_{(i=1)} = 1.577591, \quad \psi_2 = \psi_{(i=2)} = 2.965625,$$

$$\psi_3 = \psi_{(j=2)} = 2.2314888, \quad \psi_4 = \psi_{(j=1)} = 1.0607998.$$

At $k = 3$

$\Lambda_{(k)} = 878.74$, $\Lambda_{(-k)} = 1.2516477$, $\Lambda_{(n-k)} = 97.762$, $\Lambda_{(k-n)} = 1.2376084$,

$B = 1.3425443$, $P = 1748.9778$; $C = 0.0007676$, $D = 0.0057948$,

$\psi_1 = \psi_{(i=1)} = 1.0607998$, $\psi_2 = \psi_{(i=2)} = 2.2314888$,

$\psi_3 = \psi_{(i=3)} = 2.965625$, $\psi_4 = \psi_{(j=1)} = 1.577591$.

At $k = 4$

$\Lambda_{(k)} = 7818$, $\Lambda_{(-k)} = 1.252956$, $\Lambda_{(n-k)} = 9.8875$, $\Lambda_{(k-n)} = 1.1125$,

$\psi_1 = \psi_{(i=1)} = 0.5336903$, $\psi_2 = \psi_{(i=2)} = 1.147949$,

$\psi_3 = \psi_{(i=3)} = 1.7081847$, $\psi_4 = \psi_{(i=4)} = \psi_{(j=!)} = 1.706016$.

A general picture of the $\psi_{III}^{(k)}(s)$ diagram is shown in Figure 3.27

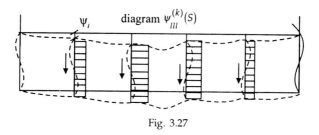

Fig. 3.27

The relation $\varphi'_{III}(s) = \psi_{III}(s)$ provides a method for constructing the $\varphi_{III}(s)$ diagram by 'integrating' the $\psi_{III}(s)$ diagram. The $\varphi_{III}^{(k)}(s)$ diagrams thus constructed for the bridge design under consideration are shown in Fig. 3.28.

3.5.4 Calculating Coefficients S_0, a_0 and Parameter β

Parameters A_i, B_i, A_j, B_j, $\tilde{A}_i$ and $\tilde{A}_j$ in eqn. (3.74) are determined respectively from eqns. (3.67), (3.69), (3.71) and (3.73) depending on the number of loaded node k.

At $k = 1$ (loading on node No. 1 of the cross-section):

$A_{(i=1)} = -2.7769869$, $B_{(i=1)} = -1.6$, $A_{(j=4)} = 0.1914169$,

$B_{(j=4,3,2,1)} = -0.4$, $A_{(j=3)} = 0.6437595$, $A_{(j=2)} = 0.6444602$,

$A_{(j=1)} = 0.7780401$, $\tilde{A}_{(i=0)} = 0.3967144$, $\tilde{A}_{(i=1)} = 0.7546302$,

$\tilde{A}_{(j=3)} = 0.1353798$, $\tilde{A}_{(j=2)} = 0.1916198$, $\tilde{A}_{(j=1)} = 0.1891975$,

$\tilde{A}_{(j=0)} = 0.1111574$, $S_0 = 0.8363076 \times 10^{-8}$.

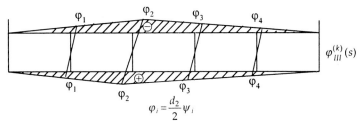

Fig. 3.28

At $k = 2$ (loading on node No. 2):

$$A_{(i=1)} = -2.3010959, \ B_{(i=1)} = -1.2, \ A_{(i=2)} = -1.8098688,$$

$$B_{(i=2)} = -1.2, \ A_{(j=3)} = 0.8345508, \ A_{(j=2)} = 1.3538872,$$

$$A_{(j=1)} = -1.5504296, \ B_{(j=0,1,2,3)} = -0.8, \ \tilde{A}_{(i=0)} = 0.3207289,$$

$$\tilde{A}_{(i=1)} = 0.529849, \ \tilde{A}_{(i=2)} = 0.25969, \ \tilde{A}_{(j=2)} = 0.3258094,$$

$$\tilde{A}_{(j=1)} = 0.371923, \ \tilde{A}_{(j=0)} = 0.2214907, \ S_0 = 0.95121948 \times 10^{-8}.$$

From symmetry:

At $k = 3$ as at $k = 2$, $S_0 = 0.95121948 \times 10^{-8}$.

At $k = 4$ as at $k = 1$, $S_0 = 0.8363076 \times 10^{-8}$.

According to eqn. (1.35), $a_0 = \int \varphi_{III}^2 (s) \times dF$, which for the bridge section under consideration gives:

$$a_0 = \left[\frac{\delta(4d_1 + d_2) d_2^2}{12} + \frac{\Delta F \cdot d_2^2}{2} \right] \sum_1^4 \psi_i^2 + \frac{\delta d_1 \, d_2^2}{6} \times \qquad \ldots (3.90)$$

$$(\psi_1 \psi_2 + \psi_2 \psi_3 + \psi_3 \psi_4).$$

At $k = 1$ and at $k = 4$, we obtain $a_0 = 1.9972976$ and at $k = 2$ and at $k = 3$, $a_0 = 18.39284$.

Finally, the values of parameter β are:

at $k = 1$ and $k = 4$, $\beta = 0.00568808$ and

at $k = 2$ and $k = 3$, $\beta = 0.00337204$.

3.5.5 Deflection Influence Surfaces

The ordinates of these surfaces are determined from eqns. (3.76) and (3.77). The results of studies of the dependence of deflection on the position of unit

108

loading for some nodal points in the midsection of the span ($z = l/2$) are given below.

1) Deflection influence surface of node No. 0 at midspan
Table 3.7 lists ordinate values at six points across and seven points along the span structure. Figure 3.29 shows the influence surface $v(l/2, 0)$. In view of symmetry, the table shows the ordinate values for half of the span. It may be noticed that the deflections of end walls (nodes) of the compartment represent the result of only the bending deformations and torsion of the bridge structure.

Table 3.7. Ordinates of the Deflection Influence Surface of Extreme Node No. 0 at Midspan ($z = 0.5\ l$ and $i = 0$)

$\dfrac{z_p}{l}$ \ k	0	1	2	3	4	5
0.0	0	0	0	0	0	0
0.2	−0.148311	−0.133875	−0.127944	−0.120500	−0.113057	−0.105613
0.4	−0.243179	−0.228489	−0.213798	−0.199108	−0.184418	−0.169728
0.5	−0.259163	−0.242408	−0.225653	−0.208898	−0.192142	−0.175387

Note: Ordinate values shown in the table should be multiplied by 10^{-7}.

2) Deflection influence surface of node No. 1 at midspan Table 3.8 gives the ordinate values and Figure 3.30 shows the influence surface $v(l/2, 1)$.

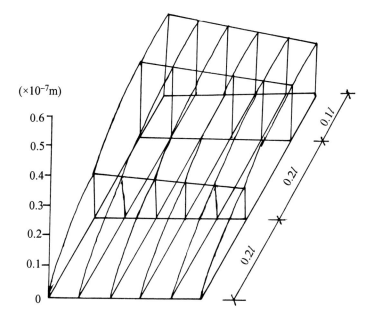

$(\times 10^{-7}\text{m})$

0.6
0.5
0.4
0.3
0.2
0.1
0

0.1 l
0.2 l
0.2 l

Fig. 3.29

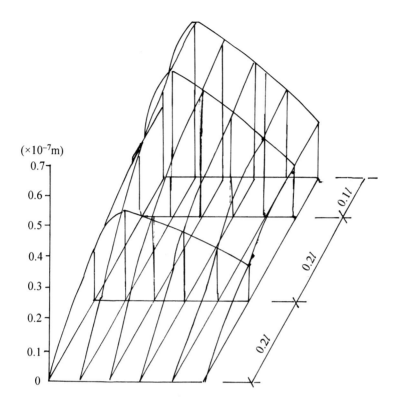

Fig. 3.30

Table 3.8. Ordinates of Deflection Influence Surface of Node No. 1 at Midspan ($z = 0.5\ l$ and $i = 1$). All Ordinates are Negative

$\dfrac{z_p}{l}$ \\ k	0	10	20	30	40	50
0.0	0	0	0	0	0	0
0.2	0.135387	0.287234	0.251347	0.209471	0.170636	0.113057
	0.135387	0.130921	0.126455	0.121989	0.117523	0.113057
0.4	0.228489	0.486219	0.421565	0.349503	0.283448	0.184418
	0.228489	0.219674	0.210860	0.202046	0.193232	0.184418
0.5	0.242408	0.514556	0.444831	0.367971	0.297624	0.192142
	0.242408	0.232354	0.232301	0.212248	0.202195	0.192142

Notes: 1) Values shown in the table should be multiplied by 10^{-7}.
2) Lower values show ordinate values ignoring section contour deformability.
3) Deflection influence surface of node No. 2 at midspan.

Table 3.9 gives the values of ordinates and Figure 3.31 shows the influence surface.

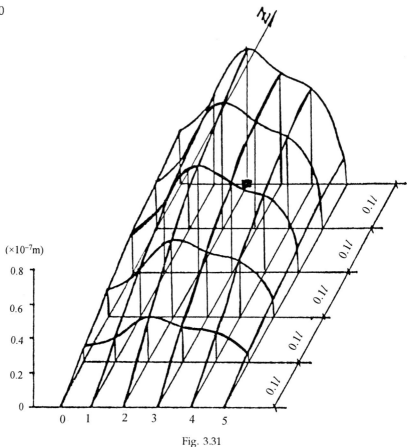

Fig. 3.31

Table 3.9. Ordinates of Deflection Influence Surface of Node No. 2 at Midspan ($z = 0.5\ l$ and $i = 2$). All Ordinates are Negative.

$\dfrac{z_p}{l}$ \ k	0	1	2	3	4	5
0.0	0	0	0	0	0	0
0.2	0.127944	0.283194	0.359537	0.307959	0.263255	0.120500
	0.127944	0.126300	0.124850	0.123400	0.121950	0.120500
0.4	0.213799	0.478131	0.603665	0.515976	0.396080	0.199708
	0.213799	0.210860	0.207922	0.204984	0.202046	0.199708
0.5	0.225653	0.505272	0.636904	0.543985	0.417493	0.208898
	0.225653	0.222302	0.218951	0.215600	0.212249	0.208898

3.5.6 Influence Surfaces for Longitudinal Normal Stresses

Ordinate values of the influence surfaces for normal stresses in the shell cross-sections of a bridge span structure are determined from eqns. (3.80). An analysis of these dependences showed that maximal normal

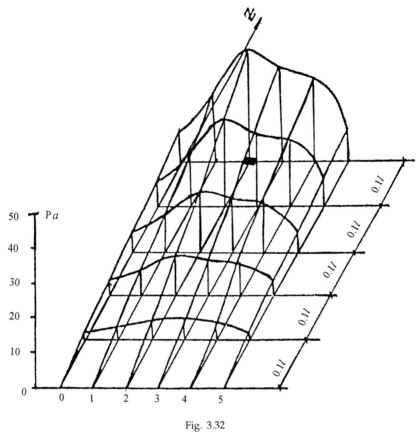

Fig. 3.32

stresses arise in the lower nodes on the central walls of the compartment at midspan.

Figure 3.32 shows a characteristic influence surface for normal stress in the lower node No. 2 on the central wall at midspan. Table 3.10 gives the values of ordinates.

For comparison, Fig. 3.33 shows the influence surface for normal stress at the same point in the span structure constructed ignoring contour deformability.

3.5.7 Influence Surfaces for Shear (Tangential) Stresses

Equations (3.83) are used for this purpose. Points on the support sections are of utmost interest in studies on shear stresses. Figsures 3.34 and 3.35 show the influence surfaces for shear stresses at nodes 0 and 1 respectively of the support section. Tables 3.11 and 3.12 show the ordinate values.

Table 3.10. Ordinates of Influence Surfaces for Normal Stress in Node No. 2 at Midspan
($z = 0.5\ l$ and $i = 2$)

$\dfrac{z_p}{l}$ ╲ k	0	1	2	3	4	5
0.0	0	0	0	0	0	0
0.2	6.628917	13.614517	17.946897	15.579999	11.874958	6.628917
	6.628917	6.628917	6.628917	6.628917	6.628917	6.628917
0.4	13.2575833	30.882593	39.074535	33.529266	26.003042	13.257833
	13.2575833	13.257833	13.257833	13.257833	13.257833	13.257833
0.5	15.467473	37.944370	47.246272	40.352993	31.515661	15.467473
	15.467473	15.467473	15.467473	15.467473	15.467473	15.467473

Note. Lower values show ordinate values ignoring section contour deformability.

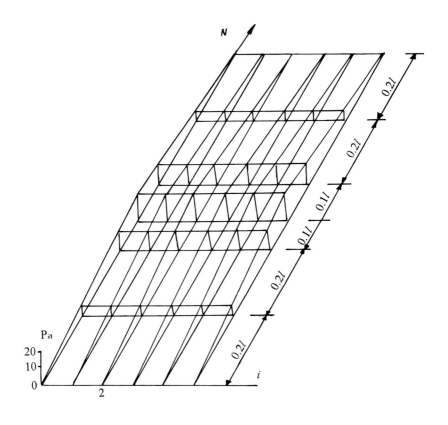

Fig. 3.33

Table 3.11. Ordinates of Influence Surfaces for Shear Stress in Node No. 0 of Support Section ($z = 0$ and $i = 0$)

$\dfrac{z_p}{l}$ \\ k	0	1	2	3	4	5
0.0	9.965250	6.623395	2.335209	−2.263221	−6.499353	−10.849105
0.2	4.735693	3.189105	0.922667	−1.200169	−3.292307	−5.442777
0.4	3.435686	2.195947	0.623802	−0.905483	−2.425973	−3.965999
0.6	2.286523	1.394340	0.390359	−0.618608	−1.629178	−2.640066
0.8	1.144351	0.675942	0.187178	−0.314372	−0.819040	−1.319922
1.0	0	0	0	0	0	0

Table 3.12. Ordinates of Influence Surfaces for Shear Stress in Node No. 1 of Support Section ($z = 0$ and $i = 1$)

$\dfrac{z_p}{l}$ \\ k	0	1	2	3	4	5
0.0	5.802378	5.910661	2.4448172	−0.563732	−3.372246	−6.686234
	5.802378	3.297766	0.801766	−1.694234	−4.190234	−6.686234
0.2	2.699999	3.031064	1.328765	−0.222452	−1.684344	−3.407083
	2.699999	1.478581	0.257165	−0.964251	−2.185667	−3.407083
0.4	1.955349	1.830833	0.796074	−0.276700	−1.336743	−2.485663
	1.955349	1.067145	0.178943	−0.709259	−1.597461	−2.485663
0.6	1.301206	1.005634	0.427707	−0.252186	−0.951871	−1.654748
	1.301206	0.710016	0.118825	−0.472366	−1.063557	−1.654748
0.8	0.650536	0.435608	0.180745	−0.148130	−0.495627	−0.827307
	0.650536	0.354969	0.059400	−0.236169	−0.531738	−0.827307
1.0	0	0	0	0	0	0

Note. Lower values show ordinate values ignoring deformability at the boundaries.

3.5.8 Influence Surfaces under Transverse Deflection Moments

Figures 3.36a and 3.36b show the influence surfaces under transverse deflection moments in the midspan in the cell upper plate sections falling respectively left and right of node No. 1. In constructing the influence surface in the section left of the node, eqns. (3.89) were used with allowance for eqns. (3.86) and (3.63). Ordinate values of this influence surface, $M(l/2, l^-)$, are given in Table 3.13.

Table 3.13. Ordinates of Influence Surface for Deflection Moment in the Upper Plate Section of the Compartment Left of Node No. 1 at Midspan

$\dfrac{z_p}{l}$ \\ k	0	1	2	3	4	5
0.0	0	0	0	0	0	0
0.2	0	−19.34432	−11.51535	−7.68230	−4.62181	0
0.4	0	−32.98586	−19.42795	−12.95054	−7.850295	0
0.5	0	−34.92348	−20.51781	−13.67487	−8.30391	0

Notes. 1) Values shown in the table should be multiplied by 10^{-5}.

2) Influence surface is symmetrical relative to $z_p/l = 0.5$.

114

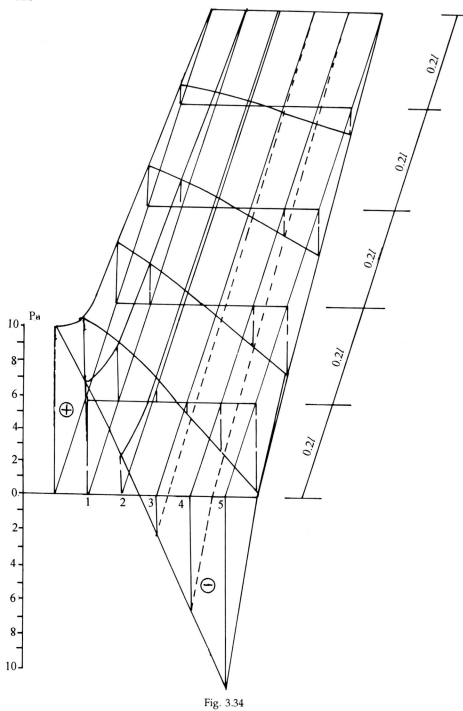

Fig. 3.34

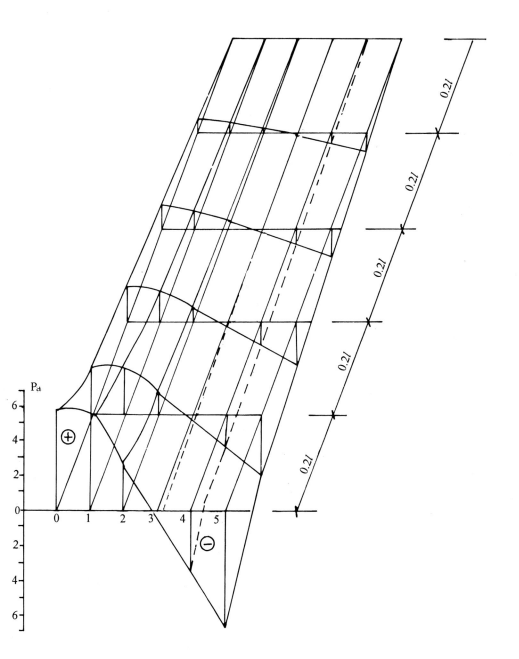

Fig. 3.35

116

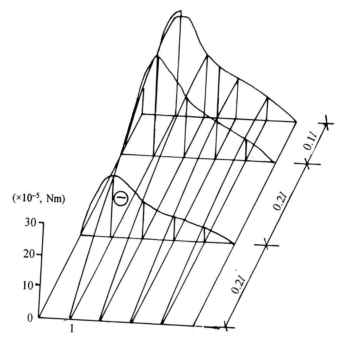

Fig. 3.36a

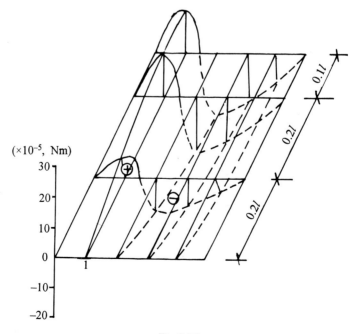

Fig. 3.36b

In constructing the influence surface for the moment in the section right of the node, eqns. (3.89) were used with allowance for eqns. (3.85) and (3.63). Ordinate values of the influence surface $M(l/2, l\,)$ are given in Table 3.14.

Table 3.14. Ordinates of Influence Surface for Deflection Moment in the Upper Plate Section of the Compartment Right of Node No. 1 at Midspan

$\dfrac{z_p}{l}$ \ k	0	1	2	3	4	5
0.0	0	0	0	0	0	0
0.2	0	8.11044	−10.37944	−8.39291	−5.21975	0
0.4	0	13.82989	−17.51107	−14.14845	−8.86592	0
0.5	0	14.64227	−18.49388	−14.93978	−9.37821	0

3.5.9 Assessing the Influence of Section Contour Deformation on Displacements, Stresses and Forces

The analyses carried out in this chapter provide not only a qualitative but also quantitative evaluation of the influence of allowing for the section contour deformation in a multiconnected span structure without diaphragms on the parameters of stressed and deformed states.

Contour deformation makes a significant contribution to the values of displacements, normal and shear stresses. Table 3.16 gives the ratios of ordinates of the deflection influence surface of node No. 1 at midspan calculated with allowance for section contour deformability (v) and without it (v^0).

Table 3.15

$\dfrac{z_p}{l}$ \ k	0	1	2	3	4	5
0.2	1	2.194	1.988	1.717	1.452	1
0.4	1	2.213	1.999	1.730	1.467	1
0.5	1	2.215	2.001	1.734	1.472	1

Table 3.16 lists similar data for the deflection of node No. 2 at midspan.

Table 3.16

$\dfrac{z_p}{l}$ \ k	0	1	2	3	4	5
0.2	1	2.242	2.880	2.496	1.937	1
0.4	1	2.268	2.903	2.517	1.960	1
0.5	1	2.273	2.909	2.523	1.967	1

Table 3.17 lists the results of comparing the values of normal stress in node No. 2 at midspan calculated with and without consideration of section contour deformability.

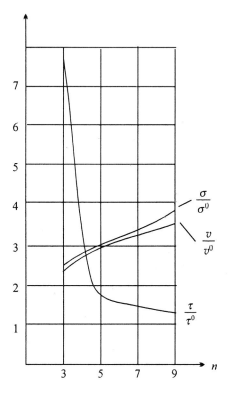

Fig. 3.37

Table 3.17

$\dfrac{z_p}{l}$ ⟍ k	0	1	2	3	4	5
0.2	1	2.054	2.707	2.350	1.791	1
0.4	1	2.329	2.947	2.511	1.961	1
0.5	1	2.453	3.055	2.609	2.038	1

Table 3.18 gives the results of comparing shear stress values of node No. 1 of the support section calculated with and without consideration of section contour deformability.

Table 3.18

$\dfrac{z_p}{l}$ ⟍ k	0	1	2	3	4	5
0.0	1	1.792	3.053	0.333	0.805	1
0.2	1	2.05	5.161	0.231	0.771	1
0.4	1	1.716	4.448	0.390	0.837	1
0.6	1	1.416	3.599	0.534	0.895	1
0.8	1	1.227	3.043	0.627	0.932	1

A quantitative analysis was carried out to understand the influence of the number of compartments in the cross-section (degree of coherence of the section) on the values of displacements, stresses and forces. Figure 3.37 illustrates the results of this analysis.

The data presented in Tables 3.15, 3.16, 3.17 and 3.18 demonstrate that the influence of contour deformability on the magnitude of displacements and normal stresses is invariably well defined, highly significant and increases with increasing extent of coherence (increasing number of compartments) of the shell. Contour deformability also influences the level of shear stresses, but with increasing number of compartments in the shell cross-section, the influence of torsion on the extent of shear stresses intensifies. This phenomenon explains the behaviour of the corresponding curve in the graph shown in Figure 3.37.

Insofar as the lateral internal forces, especially the deflection moments, are concerned, they are wholly determined by the deformability of the shell section contour.

3.5.10 *Methods for Enhancing Total Lateral Stiffness of Span Structures without Diaphragms*

Providing the sp in structure with a large number of diaphragms ensuring the contour non-deformability of a multiconnected section not only unnecessarily adds to the span weight and makes the structure expensive, but also deprives the cellular construction of its characteristic useful properties. The absence of diaphragms, however, sharply reduces the transverse stiffness of the structure.

The cellular span without diaphragms can be imparted the required level of overall transverse stiffness by various methods. Some of them are listed below.
1) Fixing the compartments alongside lateral frames.
2) Producing some members as components, e.g., upper cell wall as shown in Figure 3.38.

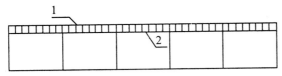

Fig. 3.38

Longitudinal ribs 1 are introduced into the conventional structure to ensure local stability and enhance the longitudinal stiffness of the upper plate of the compartment. As a result, the use of additional plate 2 is logically justified since, apart from ensuring local stability of the upper plate of the compartment, the composite cell wall invariably enhances not only the longitudinal stiffness of the upper plate of the compartment, but also the lateral stiffness of the entire span type structure.

3) Fabrication of all members of a compartment as constituent parts (Fig. 3.39).

Such a design of a cellular span type structure comprising individual components ('supercell') capable of ensuring the overall stiffness as well as the longitudinal stiffness of the upper and lower cell walls, permits transport flow at two levels.

Sokolov (1986) studied aspects of analysing cellular span type structures made up of composite members.

3.6 Analysis of Shells Comprising Two Coaxial Cylinders Connected with Circular Frames

Economic and technical requirements for structures of modern submarine systems (Bukalov and Narusbaev, 1984), tanks (Streletskii et al., 1961), panels (Kurshin, 1962), etc. necessitate increasing the structural dimensions alongwith the loads to which they are subjected.

However, such a solution is beset by several technical difficulties (complexities of stamping and welding of thick plates) and economic considerations (unjustified excessive metal consumption because of sharp reduction in strength properties of thick sheet steel).

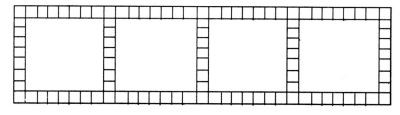

Fig. 3.39

Such difficulties at present are successfully handled in many cases by replacing continuous walls by walls constructed of composite materials which the structural layers are joined by a definite system of filler-members. Apart from high strength, stiffness, technological and economic characteristics, composite shells and panels possess several new and valuable operational properties.

In the methods available for analysing composite or the so-called three-layered shells, strips and panels, the filler is replaced in a general case by an anisotropic elastic body (Kurshin, 1962). This theory of analysis, however, has limited application (*Strength, Stability, Vibrations—A Guide*, 1968) and the results tend to be unreliable in the case of ribbed filler with comparatively widely spaced smaller number of ribs and relatively high stiffness (Volk, 1967).

In this context, it would be expedient to develop more accurate methods of analysis that permit consideration of the discrete arrangement of filler ribs. Such a consideration is comparatively simple for shells with circular ribs under axisymmetrical loadings (Sokolov, 1967, 1969, 1970b; Vol'vich and Sokolov, 1968) and in problems of cylindrical bending of three-layered plates (Sokolov, 1970a).

Solutions of more complex axisymmetrical problems are discussed below.

3.6.1 *Tank under Internal Hydrostatic Pressure (Head)*

Let us examine a vertical cylinder with two walls joined by equidistant frames subjected to internal fluid pressure (Fig. 3.40).

A solution to the problem is derived by using the well-known (Biderman, 1977; Timoshenko, 1955) analogy between a cylindrical shell during its axisymmetrical deformation and a beam on an elastic foundation. Following this analogy, let us isolate from the shell a strip of unit width by two radial sections. The branches of the elementary strip of the frame are tested along the radius toward the cylindrical axis and regarded as lateral loadings of intensity $(T_1/R_1) = (E \, \delta_1/R_1 \, R_2) \, y_1$ and $(T_2/R_2) = (E \, \delta_2/R_2^2) \, y_2$.

Moreover, the inner branch of the strip of the frame experiences the influence of given pressure $q(s)=(q_0/nd_1) \, (nd_1 - s)= (p \, nd_1/nd_1)(nd_1 - s)= p \, (nd_1 - s)$ in which p is the density of the liquid filling the tank.

The differential equations for the bending of the elementary frame will then be:

$$D_1 \, y_1^{IV} + \frac{E \, \delta_1}{R_1 \, R_2} \, y_1 = q(s) \text{—for the inner branch, and} \qquad \text{... (3.91)}$$

$$D_2 \, y_2^{IV} + \frac{E \, \delta_2}{R_2^2} \, y_2 = 0 \text{—for the outer branch.} \qquad \text{... (3.92)}$$

The above equations are wholly identical to the equation for beam bending on an elastic foundation. Thus, a unit strip isolated from the shell can be considered a two-branched frame whose branches are joined with the elastic foundations substituting for the effect of curved fibres and the nodes with the elastic radial supports substituting for the effect of circular frames. The analytical scheme thus obtained is shown in Fig. 3.41. The stiffness parameters of the members of a strip of frame and elastic foundation are determined by the following relationships:

$$D_1 = \frac{E \, \delta_1^3 \, R_1}{12(1 - v^2)R_2}, \quad D_2 = \frac{E \, \delta_2^3}{12(1 - v^2)} \text{—flexural stiffness of branches,}$$

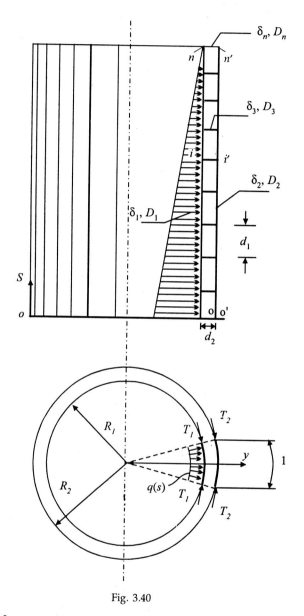

Fig. 3.40

$$D_3 = \frac{E\delta_3^3}{12(1-v^2)} \cdot \frac{R_1 + R_2}{2R_2} \text{ —flexural stiffness of intermediate layers,}$$

$$D_n = \frac{E\delta_n^3}{12(1-v^2)} \cdot \frac{R_1 + R_2}{2R_2} \text{ —flexural stiffness of upper layers,}$$

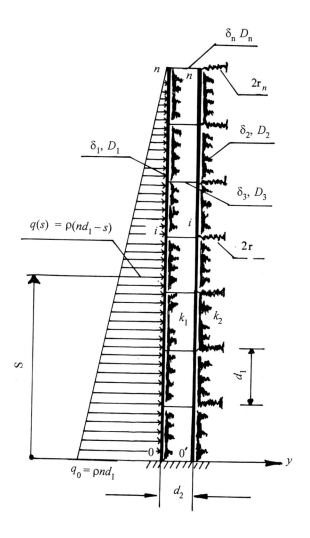

Fig. 3.41

$$2r = \frac{EF_{III}}{R_0^2} = \frac{E\delta_3 4d_2}{(R_1 + R_2)^2} = \frac{4E\delta_3 (R_2 - R_1)}{(R_1 + R_2)^2} \text{ —stiffness of intermediate}$$

elastic supports,

$$2r_n = \frac{4E\delta_0 (R_2 - R_1)}{(R_1 + R_2)^2} \text{ —stiffness of upper elastic support,}$$

$$k_1 = \frac{E\delta_1}{R_1 R_2}$$ —stiffness of elastic foundation of inner branch and

$$k_2 = \frac{E\delta_2}{R_2^2}$$ —stiffness of elastic foundation of outer branch.

The system shown in Fig. 3.41 represents a regular one and permits solution in a closed analytical form. For this purpose, the system of equations in finite differences and the displacement method in a developed form can be applied. The angular (ω_i, ω_i') and linear radial (ψ_i) nodal displacements are used as unknowns of the displacement method. Unlike the solutions derived by Sokolov (1967), the simplifying assumption about infinitely high flexural stiffness of frames as well as consideration of longitudinal (along axis S) deformability of the branches of the strip have been ignored. Forces along the ends of elements joined with the elastic foundation caused by displacements of the ends of elements (Fig. 3.42) are determined using equations suggested by Chudnovskii (1952).

$$M_{ab} = \frac{D_1}{d_1}\left(\alpha_{ab}\,\omega_a + \beta_{ab}\cdot\omega_b - \gamma_{ab}\,\frac{\psi_a}{d_1} - \eta_{ab}\,\frac{\psi_b}{d_1} \right),$$

$$Q_{ab} = -\frac{D_1}{d_1^2}\left(\gamma_{ab}\,\omega_a + \eta_{ab}\,\omega_b - \mu_{ab}\,\frac{\psi_a}{d_1} - \zeta_{ab}\,\frac{\psi_b}{d_1} \right), \qquad \dots (3.93)$$

$$M_{ba} = \frac{D_1}{d_1}\left(\alpha_{ba}\,\omega_b + \beta_{ba}\cdot\omega_a - \gamma_{ba}\,\frac{\psi_b}{d_1} - \eta_{ba}\,\frac{\psi_a}{d_1} \right),$$

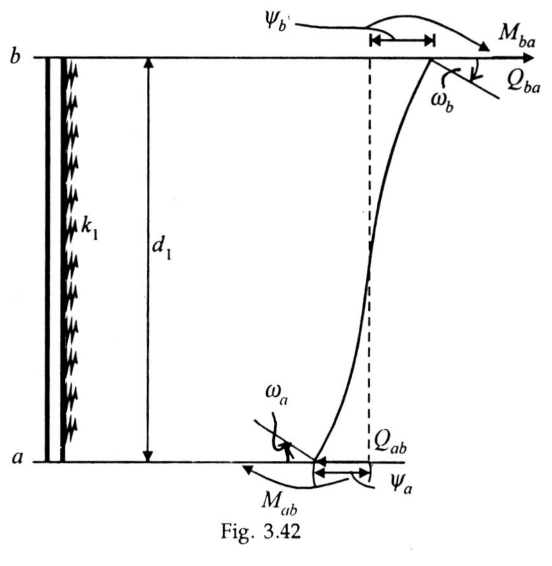

Fig. 3.42

$$Q_{ba} = \frac{D}{d_1^2}\left(\gamma_{ba}\,\omega_b + \eta_{ba}\,\omega_a - \mu_{ba}\,\frac{\psi_b}{d_1} - \zeta_{ba}\,\frac{\psi_a}{d_1}\right).$$

Special functions $\alpha_{ab} = \alpha_{ba}$, $\beta_{ab} = \beta_{ba}$, $\gamma_{ab} = \gamma_{ba}$, $\eta_{ab} = \eta_{ba}$, $\mu_{ab} = \mu_{ba}$ and $\zeta_{ab} = \zeta_{ba}$ are derived from equations given by Chudnovskii as a partial case in the absence of longitudinal force. Equation (3.91) is rewritten as follows:

$$y_1^{IV} + 4u^4 \cdot y_1 = \frac{q(s)}{D_1},$$

in which

$$u = \sqrt[4]{\frac{k_1}{4D_1}}. \qquad \qquad \text{... (3.94)}$$

The characteristic equation corresponding to the homogeneous equation will be:

$$\lambda^4 + 4u^4 = 0 \text{ and its root}$$

$$\lambda_1 = -2u^2 i,\ \lambda_2 = 2u^2 i,\ i = \sqrt{-1}. \qquad \text{... (3.95)}$$

On considering eqn. (3.95), Chudnovskii's special functions assume a simple form:

$$\alpha_{ab} = \alpha_{ba} = u \cdot \frac{\sin 2u - sh2u}{\sin^2 u - sh^2 u},$$

$$\beta_{ab} = \beta_{ba} = 2u \cdot \frac{shu \cdot \cos u - chu \cdot \sin u}{\sin^2 u - sh^2 u},$$

$$\gamma_{ab} = \gamma_{ba} = -2u^2 \cdot \frac{sh^2 u \cdot \cos^2 u + ch^2 u \cdot \sin^2 u}{\sin^2 u - sh^2 u},$$

$$\eta_{ab} = \eta_{ba} = +4u^2 \cdot \frac{shu \cdot \sin u}{\sin^2 u - sh^2 u}, \qquad \text{... (3.96)}$$

$$\mu_{ab} = \mu_{ba} = -2u^3 \cdot \frac{sh2u \cdot \sin 2u}{\sin^2 u - sh^2 u},$$

$$\zeta_{ab} = \zeta_{ba} = -4u^3 \cdot \frac{shu \cdot \cos u + chu \cdot \sin u}{\sin^2 u - sh^2 u}.$$

To derive equations for nodal moments and lateral forces caused by a given loading, let us resolve the auxiliary problem (Fig. 3.43) by the method of initial parameters.

Let us express the known equations for the derivatives of bending moments along abscissas:

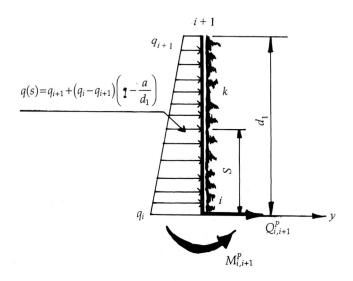

Fig. 3.43

$$\frac{dM}{ds} = Q, \quad \frac{d^2 M}{ds^2} = q_i + (q_{i-1} - q_i)\, s + k \cdot y \cdot l,$$

$$\frac{d^3 M}{ds^3} = q_{i-1} - q_i + k \cdot y' \cdot l, \qquad \frac{d^4 M}{ds^4} = k \cdot y'' \cdot l.$$

Considering that $y'' = -(M/D)$, the latter equation can be assigned the form:

$$\frac{d^4 M}{ds^4} + 4u^4 \cdot M = 0. \qquad \qquad \text{... (3.97)}$$

The general solution for the homogeneous equation is:

$$M = C_1 \sin \bar{s} \cdot sh\,\bar{s} + C_2 \sin \bar{s} \cdot ch\,\bar{s} + C_3 \cos \bar{s} \cdot sh\,\bar{s} + C_4 \cos \bar{s} \cdot ch\,\bar{s} \quad \text{... (3.98)}$$

in which

$$\bar{s} = u \cdot s.$$

The integration constants are expressed in terms of initial parameters as follows:

$$C_1 = \frac{1}{2} U_0, \; C_2 = \frac{1}{2} V_0 + \frac{1}{4} W_0, \; C_3 = \frac{1}{2} V_0 - \frac{1}{4} W_0, \; C_4 = M_0,$$

in which

$$V_0 = \frac{Q_0}{u}, \; U_0 = \frac{q_i + k \cdot y_0 \cdot l}{u^2},$$

$$W_0 = \frac{q_{i-1} - q_i + k \cdot y_0' \cdot l}{u^3}. \qquad \qquad \text{... (3.99)}$$

Taking into account the notation of A.N. Krylov for the basic functions of beams, eqn. (3.98) assumes the form

$$M = M_0 \tilde{A} + V_0 \tilde{B} + U_0 \tilde{C} + W_0 \tilde{D}. \qquad \ldots(3.100)$$

in which

$$\tilde{A} = \cos \bar{s} \cdot ch \bar{s}, \quad \tilde{B} = \frac{1}{2}\left(\sin \bar{s} \cdot ch \bar{s} + \cos \bar{s} \cdot sh \bar{s}\right), \qquad \ldots(3.101)$$

$$\tilde{C} = \frac{1}{2}\sin \bar{s} \cdot sh \bar{s}, \quad \tilde{D} = \frac{1}{4}\left(\sin \bar{s} \cdot ch \bar{s} - \cos \bar{s} \cdot sh \bar{s}\right).$$

Then, the following equations are available for determining the moments, sheer forces, linear and angular displacements in an arbitrary beam section:

$$M = M_0 \cdot \tilde{A} + V_0 \tilde{B} + U_0 \tilde{C} + W_0 \cdot \tilde{D},$$

$$V = -4M_0 \tilde{D} + V_0 \tilde{A} + U_0 \tilde{B} + W_0 \tilde{C}, \qquad \ldots(3.102)$$

$$U = -4M_0 \tilde{C} - 4V_0 \tilde{D} + U_0 \tilde{A} + W_0 \tilde{B},$$

$$W = -4M_0 \tilde{B} - 4V_0 \tilde{C} - 4U_0 \tilde{D} + W_0 \tilde{A},$$

in which

$$V = \frac{Q}{u}, \quad U = \frac{q_i + \left(q_{i-1} - q_i\right)\bar{s} + k \cdot y \cdot l}{u^2},$$

$$W = \frac{q_{i-1} - q_i + k \cdot y' \cdot l}{u^3}. \qquad \ldots(3.103)$$

In the scheme under consideration (Fig. 3.44), the moment and shear force in node i coinciding with the initial parameters M_0 and Q_0 are of interest.

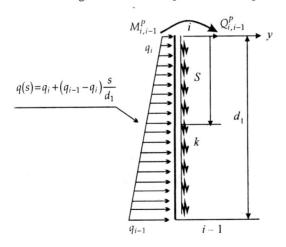

Fig. 3.44

By subjecting eqn. (3.102) to the limiting conditions:

$$\bar{s} = 0; \qquad\qquad y = 0,\ y' = 0,$$
$$\bar{s} = ud_1; \qquad\qquad y = 0,\ y' = 0,$$

we find:

$$U_0 = \frac{q_i}{u_1^2},\quad W_0 = \frac{q_{i-1} - q_i}{d_1\, u_1^3},$$

$$-4M_0\tilde{C} - 4V_0\tilde{D} + \frac{q_i}{u_1^2}\,\tilde{A} + \frac{q_{i-1} - q_i}{d_1\, u_1^3}\,\tilde{B} = 0,$$

$$-4M_0\tilde{B} - 4V_0\tilde{C} - 4\frac{q_i}{u_1^2}\,\tilde{D} + \frac{q_{i+1} - q_i}{d_1\, u_1^3}\,\tilde{A} = 0.$$

Resolving the resultant equations system, we find:

$$M_0 = \frac{1}{4\,u_1^4}\left(q_1\cdot G_I + \frac{q_{i+1} - q_i}{d_1}\,A_I\right),\quad V_0 = -\frac{1}{4\,u_1^4}\left(q_1 M_I + \frac{q_{i-1} - q_i}{d_1}\,G_I\right).$$

Then,

$$M_{i,i-1}^P = M_0 = \frac{1}{4\,u_1^4}\left(q_i\cdot G_I + \frac{q_{i-1} - q_i}{d_1}\,A_I\right), \qquad\qquad \text{... (3.104)}$$

$$Q_{i,i-1}^P = u\cdot V_0 = \frac{1}{4\,u_1^3}\left(q_i\cdot M_I + \frac{q_{i-1} - q_i}{d_1}\,G_I\right),$$

in which

$$A_I = u_1 d_1\,\frac{\sin 2u_1 d_1 - sh\,2u_1 d_1}{\sin^2 u_1 d_1 - sh^2 u_1 d_1}, \qquad\qquad \text{... (3.105)}$$

$$G_I = -2u_1^2\, d_1^2\cdot\frac{sh^2\, u_1 d_1\cdot\cos^2 u_1 d_1 + ch^2\, u_1 d_1\cdot\sin^2 u_1 d_1}{\sin^2 u_1 d_1 - sh^2 u_1 d_1}$$

$$M_I = -2u_1^3\, d_1^3\cdot\frac{sh\,2u_1 d_1 + \sin 2u_1 d_1}{\sin^2 u_1 d_1 - sh^2 u_1 d_1}.$$

Equations for the reactive forces in node i due to loading of the $i + 1$st panel (Fig. 3.43) are derived similarly and assume the form:

$$M_{i,i+1}^P = \frac{1}{4\,u_1^4}\left(q_i\cdot G_I + \frac{q_{i+1} - q_i}{d_1}\,A_I\right),$$

$$Q_{i,i+1}^P = \frac{1}{4\,u_1^3}\left(q_i\cdot M_I + \frac{q_{i+1} - q_i}{d_1}\,G_I\right). \qquad\qquad \text{... (3.106)}$$

In eqns. (3.105) and (3.106):

$$q_i = \frac{q_0}{n}(n-i) = \rho d_1(n-i), q_{i-1} = \rho d_1(n-i+1), \ q_{i+1} = \rho d_1(n-i-1).$$

Let us examine the equilibrium of the i-th node of the strip of the frame in the deformed state (Fig. 3.45):

$$\sum M_i = M_{i,i-1} + M_{i,i+1} + M_{i,i'} + M^P_{i,i-1} + M^P_{i,i+1} = 0.$$

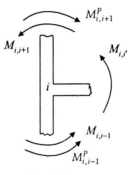

Fig. 3.45

Using equations (3.93), (3.105) and (3.106) for $M_{i,i-1}$, $M_{i,i+1}$, $M^P_{i,i-1}$ and $M^P_{i,i+1}$ and the usual equation of the displacement method in a developed form for $M_{i,i'}$, the equilibrium equation for the i-th node is derived as follows:

$$\frac{D_1}{d_1}\left(\alpha_1\omega_i + \beta_1\omega_{i-1} - \gamma_1\frac{\psi_i}{d_1} + \eta_1\frac{\psi_{i-1}}{d_1}\right) + \frac{D_1}{d_1} \times$$

$$\left(\alpha_1\omega_i + \beta_1\omega_{i+1} + \gamma_1\frac{\psi_i}{d_1} - \eta_1\frac{\psi_{i+1}}{d_1}\right) + \frac{2D_3}{d_2}\left(2\omega_i + \omega_{i'}\right) +$$

$$\frac{1}{4u_1^4}\cdot\frac{q_{i-1} - q_{i+1}}{d_1}A_1 = 0,$$

or, after simple conversions:

$$2\left(\alpha_1 + 2\frac{D_3\,d_1}{D_1\,d_2}\right)\omega_i + \beta_1\left(\omega_{i-1} + \omega_{i+1}\right) + 2\frac{D_3\,d_1}{D_1\,d_2}\omega_{i'} +$$

$$\frac{\eta_1}{d_1}\left(\psi_{i-1} - \psi_{i+1}\right) = -2\frac{\rho d_1}{k_1}. \tag{a}$$

The equilibrium equation for the i'-th node will be (Fig. 3.46):

$$\sum M_{i'} = M_{i',j'-1} + M_{i',i'+1} + M_{i',i'} = 0.$$

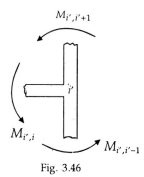

Fig. 3.46

Taking into account equation (3.93), we derive:

$$2\left(\alpha_2 + 2\frac{D_3}{D_2}\frac{d_1}{d_2}\right)\omega_{i'} + \beta_2\left(\omega_{i'-1} + \omega_{i'+1}\right) + 2\frac{D_3}{D_2}\frac{d_1}{d_2}\,\omega_i +$$

$$\frac{\eta_2}{d_1}\left(\psi_{i-1} - \psi_{i+1}\right) = 0. \tag{b}$$

Let us consider the sum of the projections of forces acting on the i-th layer as the third equation of statics on the horizontal axis (Fig. 3.47):

$$\Sigma F_y = Q_{i,i+1} - Q_{i,i-1} + Q_{i',i'-1} - Q_{i',i'-1} - Q_{i,i+1}^P - 2r\psi_i - Q_{i,i-1}^P = 0.$$

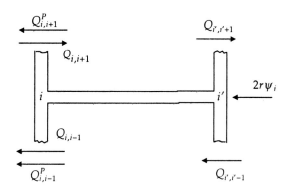

Fig. 3.47

Considering equations (3.93), (3.105) and (3.106), the above equation assumes the form:

$$\eta_1\left(\omega_{i-1}-\omega_{i+1}\right)+\frac{D_2}{D_1}\,\eta_2\left(\omega_{i'-1}-\omega_{i'+1}\right)-\frac{2}{d_1}\left(\mu_1+\frac{D_2}{D_1}\mu_2\right)\psi_l+\frac{1}{d_1}\times$$

$$\left(\zeta_1+\frac{D_2}{D_1}\zeta_1\right)\left(\psi_{i-1}+\psi_{i+1}\right)-2r\psi_i\cdot\frac{d_1^3}{D_1}=-2u_1\,\frac{p}{k_1}\,M_1\cdot(n-i). \qquad \text{(c)}$$

Equations (a), (b) and (c) form a system in finite differences

$$2\left(\alpha_1+2\frac{D_3\,d_1}{D_1\,d_2}\right)\omega_i+\beta_1\left(\omega_{i-1}+\omega_{i+1}\right)+2\frac{D_3\,d_1}{D_1\,d_2}\,\omega_{i'}+\frac{\eta_1}{d_1}\times$$

$$\left(\psi_{i-1}-\psi_{i+1}\right)=-2\frac{\rho d_1}{k_1},$$

$$2\left(\alpha_2+2\frac{D_3\,d_1}{D_1\,d_2}\right)\omega_{i'}+\beta_2\left(\omega_{i'-1}+\omega_{i'+1}\right)+2\frac{D_3\,d_3}{D_2\,d_2}\,\omega_i+\frac{\eta_2}{d_1}\times$$

$$\left(\psi_{i-1}-\psi_{i+1}\right)=0, \qquad\qquad \text{... (3.107)}$$

$$\eta_1\left(\omega_{i-1}-\omega_{i+1}\right)+\frac{D_2}{D_1}\,\eta_2\left(\omega_{i'-1}-\omega_{i'+1}\right)-\left[\frac{2}{d_1}\left(\mu_1+\frac{D_2}{D_1}\mu_2\right)+2r\frac{d_1^3}{D_1}\right]\psi_i+$$

$$\frac{1}{d_1}\left(\zeta_1+\frac{D_2}{D_1}\zeta_2\right)\left(\psi_{i-1}+\psi_{i+1}\right)=-2u_1\,\frac{p}{k_1}\,M_1\,(n-i),$$

in which α_1, β_1, γ_1, η_1, μ_1 and ζ_1 pertain to a member of the inner shell and α_2, β_2, γ_2, η_2, μ_2 and ζ_2 of the outer shell.

According to Bleikh and Melan (1938), the general solution for the corresponding homogeneous equations is found in the form:

$$\omega_i^0=\lambda^i,\quad \omega_{i'}^0=\varepsilon\lambda^i,\quad \psi_i^0=t\lambda^i \qquad\qquad \text{... (3.108)}$$

in which ε and t represent unknown parameters.

After substituting eqns. (3.108) in (3.107) and equating the main determinant to zero, we obtain the following characteristic equation:

$$\lambda^6+A_0\,\lambda^5+B_0\,\lambda^4+C_0\,\lambda^3+B_0\,\lambda^2+A_0\,\lambda+1=0, \qquad \text{... (3.109)}$$

in which

$$A_0=2\left(a_1\beta_2+a_2\,\beta_1\right)\left(\zeta_1+\frac{D_2}{D_1}\zeta_2\right)-\left(a_2\,\eta_1^2+\alpha_1\,\frac{D_2}{D_1}\,\eta_2^2\right)-$$

$$\beta_1\beta_2\left(\mu_1+\mu_2\frac{D_2}{D_1}+r\frac{d_1^3}{D_1}\right)+2\frac{D_3}{D_1}\cdot\frac{d_1}{d_2}\,\eta_1\,\eta_2 \,\Big/$$

$$\beta_1\,\beta_2\left(\xi_1+\frac{D_2}{D_1}\,\xi_2\right)-\left(\eta_1^2\,\beta_2+\frac{D_2}{D_1}\,\eta_2^2\,\beta_1\right),$$

$$B_0 = \left(4a_1a_2 + 3\beta_1\beta_2 - 4\frac{D_3}{D_1}\frac{D_3}{D_2}\frac{d_1^2}{d_2^2}\right)\left(\zeta_1 + \frac{D_2}{D_1}\zeta_2\right) - 4\left(a_1\beta_2 + a_2\beta_1\right) \times$$

$$\left(\mu_1 + \frac{D_2}{D_1}\mu_2 + r\frac{d_1^3}{D_1}\right) + \eta^2\beta_2 + \frac{D_2}{D_1}\cdot\eta_2^2\beta_1 \Bigg/$$

$$\beta_1\beta_2\left(\xi_1 + \frac{D_2}{D_1}\xi_2\right) - \left(\eta_1^2\beta_2 + \frac{D_2}{D_1}\eta_2^2\beta_1\right),$$

$$C_0 = 4\frac{(a_1\beta_2 + a_2\beta_1)\left(\zeta_1 + \frac{D_2}{D_1}\zeta_2\right) - \left(2a_1a_2 + \beta_1\beta_2 - 2\frac{D_3^2}{D_1 D_2}\frac{d_1^2}{d_2^2}\right)\left(\mu_1 + \frac{D_2}{D_1}\mu_2 + r\frac{d_1^3}{D_1}\right)}{\beta_1\beta_2\left(\xi_1 + \frac{D_2}{D_1}\xi_2\right) - \left(\eta_1^2\beta_2 + \frac{D_2}{D_1}\eta_2^2\beta_1\right)} +$$

$$\frac{\left(a_2\eta_1^2 + a_1\frac{D_2}{D_1}\eta_2^2\right) - 2\frac{D_3}{D_1}\frac{d_1}{d_2}\eta_1\eta_2}{\beta_1\beta_2\left(\xi_1 + \frac{D_2}{D_1}\xi_2\right) - \left(\eta_1^2\beta_2 + \frac{D_2}{D_1}\eta_2^2\beta_1\right)}, \qquad \text{... (3.110)}$$

$$a_1 = \alpha_1 + 2\frac{D_3}{D_1}\frac{d_1}{d_2}, \quad a_2 = \alpha_2 + 2\frac{D_3}{D_2}\cdot\frac{d_1}{d_2}.$$

Equation (3.109) is represented as follows:

$$\left(\lambda^2 + v_1\lambda + 1\right)\left(\lambda^2 + v_2\lambda + 1\right)\left(\lambda^2 + v_3\lambda + 1\right) = 0,$$

in which v_1, v_2 and v_3 are the cube roots of the equation:

$$v^3 - A_0 v_2 + \left(B_0 - 3\right)v + 2A_0 - C_0 = 0. \qquad \text{... (3.111)}$$

By substituting $y = v - (A_0/3)$, the above equation is transformed into:

$$y^3 + 3\overline{p}\,y + 2\overline{q} = 0, \text{ in which}$$

$$3\overline{p} = B_0 - 3 - \frac{1}{3}A_0^2, \; 2\overline{q} = -\frac{2}{27}A_0^3 + \frac{A_0\left(B_0 - 3\right)}{3} + 2A_0 - C_0. \quad \text{... (3.112)}$$

It is known that at $\overline{q}^2 + \overline{p}^3 \leq 0$:

$$y_1 = 2r\cos\frac{\varphi}{3}, \; y_2 = 2r\cos\left(60° - \frac{\varphi}{3}\right), \; y_3 = 2r\cos\left(60° + \frac{\varphi}{3}\right), \text{ in which}$$

$$\cos\varphi = \frac{\overline{q}}{r^3}, \; r = \pm\sqrt{|\overline{p}|} \text{ (sign should coincide with that of } \overline{q}\text{)}.$$

Then,

$$v_1 = y_1 + \frac{A_0}{3}, \quad v_2 = y_2 + \frac{A_0}{3}, \quad v_3 = y_3 + \frac{A_0}{3}. \qquad \dots \text{(3.113)}$$

The roots of equation (3.109) corresponding to the three values of v will be:

$$\lambda_1 = -\frac{v_1}{2} + \frac{1}{2}\sqrt{v_1^2 - 4}, \; \frac{1}{\lambda_1},$$

$$\lambda_2 = -\frac{v_2}{2} + \frac{1}{2}\sqrt{v_2^2 - 4}, \; \frac{1}{\lambda_2}, \qquad \dots \text{(3.114)}$$

$$\lambda_3 = -\frac{v_3}{2} + \frac{1}{2}\sqrt{v_1^2 - 4}, \; \frac{1}{\lambda_3}.$$

To determine parameters ε and t, let us solve simultaneously the first and third equilibrium eqns. (3.107) without the right-hand parts after substituting eqn. (3.108) in them and reducing by λ^{i-1}:

$$2\frac{D_3}{D_1}\frac{d_1}{d_2}\,\lambda\cdot\varepsilon + \frac{\eta_1}{d_1}\left(1 - \lambda^2\right)t = -\left[2a_1\lambda + \beta_1\left(1 + \lambda^2\right)\right],$$

$$\frac{D_2}{D_1}\eta_2\left(1 - \lambda^2\right)\varepsilon + \left\{-\left[\frac{2}{d_1}\left(\mu_1 + \frac{D_2}{D_1}\mu_2\right) + 2r\frac{d_1^3}{D_1}\right]\lambda + \right.$$

$$\left.\frac{1}{d_1}\left(\zeta_1 + \frac{D_2}{D_1}\xi_2\right)\left(1 - \lambda^2\right)\right\}t = -\eta_1\left(1 - \lambda^2\right). \qquad \dots \text{(3.115)}$$

hence

$$\varepsilon = -\eta_1^2\left(1 - \lambda^2\right)^2 - \left[2a_1\left(\lambda - \lambda^3\right) + \beta_1\left(1 - \lambda^4\right)\right]\left(\zeta_1 + \frac{D_2}{D_1}\zeta_2\right) +$$

$$2\left[2a_1\lambda^2 + \beta_1\left(\lambda^3 + \lambda\right)\right]\left(\mu_1 + \frac{D_2}{D_1}\mu_2 + r\frac{d_1^4}{D_1}\right) \Bigg/$$

$$\eta_1\eta_2\frac{D_2}{D_1}\left(1 - \lambda^2\right)^2 + 4\frac{D_3}{D_1}\frac{d_1}{d_2}\left(\mu_1 + \frac{D_2}{D_1}\mu_2 + r\frac{d_1^4}{D_1}\right)\lambda^2 -$$

$$2\frac{D_3}{D_1}\frac{d_1}{d_2}\left(\zeta_1 + \frac{D_2}{D_1}\cdot\zeta_2\right)\left(\lambda - \lambda^3\right),$$

$$t = -d_1 \cdot 2\eta_1 \left(a_1 + \frac{D_3}{D_2} \frac{d_1}{d_2} \right) (\lambda - \lambda^3) + \beta_1 \eta_2 \left(1 - \lambda^4 \right) \Big/$$

$$\eta_1 \eta_2 \frac{D_2}{D_1} \left(1 - \lambda^2 \right)^2 + 4 \frac{D_3}{D_1} \frac{d_1}{d_2} \left(\mu_1 + \frac{D_2}{D_1} \mu_2 + r \frac{d_1^4}{D_1} \right) \lambda^2 -$$

$$2 \frac{D_3}{D_1} \frac{d_1}{d_2} \left(\zeta_1 + \frac{D_2}{D_1} \cdot \zeta_2 \right) (\lambda - \lambda^3). \qquad \text{... (3.116)}$$

Six values of ε and t are determined corresponding to the six values of λ. Further, it is quite clear that:

$$\varepsilon(\lambda_1) = \varepsilon\left(\frac{1}{\lambda_1}\right) = \varepsilon_1, \qquad t(\lambda_1) = t\left(\frac{1}{\lambda_1}\right) = t_1,$$

$$\varepsilon(\lambda_2) = \varepsilon\left(\frac{1}{\lambda_2}\right) = \varepsilon_2, \qquad t(\lambda_2) = t\left(\frac{1}{\lambda_2}\right) = t_2,$$

$$\varepsilon(\lambda_3) = \varepsilon\left(\frac{1}{\lambda_3}\right) = \varepsilon_3, \qquad t(\lambda_3) = t\left(\frac{1}{\lambda_3}\right) = t_3.$$

Then, the general solution for the system of eqns. (3.107) will be

$$\omega_i^0 = C_1 \lambda_1^i + C_2 \lambda_1^{-i} + C_3 \lambda_2^i + C_4 \lambda_2^{-i} + C_5 \lambda_3^i + C_6 \lambda_3^{-i},$$

$$\omega_{i'}^0 = \varepsilon_1 \left(C_1 \lambda_1^i + C_2 \lambda_1^{-i} \right) + \varepsilon_2 \left(C_3 \lambda_2^i + C_4 \lambda_2^{-i} \right) + \varepsilon_3 \left(C_5 \lambda_3^i + C_6 \lambda_3^{-i} \right),$$

$$\psi_i^0 = t_1 \left(C_1 \lambda_1^i + C_2 \lambda_1^{-i} \right) + t_2 \left(C_3 \lambda_2^i + C_4 \lambda_2^{-i} \right) + t_3 \left(C_5 \lambda_3^i + C_6 \lambda_3^{-i} \right).$$

$$\text{... (3.117)}$$

Conforming to the structure of the right-hand parts of eqns. (3.107), their partial solution is found in the following form:

$$\omega_i^{00} = \rho(A_1 + B_1 \cdot i) + \rho(A_2 + B_2 \cdot i) + \rho(A_3 + B_3 \cdot i),$$

$$\omega_{i'}^{00} = \rho\varepsilon_1 (A_1 + B_1 \cdot i) + \rho\varepsilon_2 (A_2 + B_2 \cdot i) + \rho\varepsilon_3 (A_3 + B_3 \cdot i), \qquad \text{... (3.118)}$$

$$\psi_i^{00} = \rho t_1 (A_1 + B_1 \cdot i) + \rho t_2 (A_2 + B_2 \cdot i) + \rho t_3 (A_3 + B_3 \cdot i).$$

After substituting in eqn. (3.107) and equating the coefficients for identical degrees of i, we find:

$$\omega_i^{00} = -\frac{\rho d_1}{k_1} \left\{ \frac{1}{a_1 + \beta_1 + \dfrac{D_3}{D_1} \dfrac{d_1}{d_2} \varepsilon_1} \cdot -\frac{u_1 M_1}{d_1 t_3^2} \times \right.$$

$$\frac{d_1\left(\eta_1 + \dfrac{D_2}{D_1}\eta_2\,\varepsilon_3\right) - t_3\left(\zeta_1 + \dfrac{D_2}{D_1}\zeta_2 - \mu_1 - \dfrac{D_2}{D_1}\mu_2 - r\dfrac{d_1^3}{D_1}\right)(n-i)}{\left(\zeta_1 + \dfrac{D_2}{D_1}\zeta_2 - \mu_1 - \dfrac{D_2}{D_1}\mu_2 - \gamma\dfrac{d_1^3}{D_1}\right)^2}\Bigg\},$$

$$\omega_i^{00} = -\frac{\rho d_1}{k_1}\left\{\frac{\varepsilon_1}{a_1 + \beta_1 + \dfrac{D_3}{D_1}\dfrac{d_1}{d_2}\varepsilon_1} - \frac{u_1 M_1 \varepsilon_3}{d_1 t_3^2}\times\right.$$

$$\frac{d_1\left(\eta_1 + \dfrac{D_2}{D_2}\eta_2\,\varepsilon_3\right) - t_3\left(\zeta_1 + \dfrac{D_2}{D_2}\zeta_2 - \mu_1 - \dfrac{D_2}{D_2}\mu_2 - r\dfrac{d_1^3}{D_1}\right)(n-i)}{\left(\zeta_1 + \dfrac{D_2}{D_2}\zeta_2 - \mu_1 - \dfrac{D_2}{D_2}\mu_2 - \gamma\dfrac{d_1^3}{D_1}\right)^2}\Bigg\},$$

$$\psi_i^{00} = -\frac{\rho d_1}{k_1}\left\{\frac{t_1}{a_1 + \beta_1 + \dfrac{D_3}{D_1}\dfrac{d_1}{d_2}\varepsilon_1} - \frac{u_1 M_1}{d_1 t_3^2}\times\right.$$

$$\left.\frac{d_1\left(\eta_1 + \dfrac{D_2}{D_1}\eta_2\,\varepsilon_3\right) - t_3\left(\zeta_1 + \dfrac{D_2}{D_1}\zeta_2 - \mu_1 - \dfrac{D_2}{D_1}\mu_2 - r\dfrac{d_1^3}{D_1}\right)(n-i)}{\left(\zeta_1 + \dfrac{D_2}{D_1}\zeta_2 - \mu_1 - \dfrac{D_2}{D_1}\mu_2 - \gamma\dfrac{d_1^3}{D_1}\right)^2}\right\}.$$

Let us introduce the notation:

$$L = \zeta_1 + \frac{D_2}{D_1}\zeta_2 - \mu_1 - \frac{D_2}{D_1}\mu_2 - \gamma\frac{d_1^3}{D_1}. \qquad \dots (3.119)$$

Then, the partial solutions are expressed more compactly as:

$$\omega_i^{00} = -\frac{\rho d_1}{k_1}\left[\frac{1}{a_1 + \beta_1 + \dfrac{D_3}{D_1}\dfrac{d_1}{d_2}\varepsilon_1} - \frac{u_1 M_1}{d_1 t_3^2}\cdot\frac{d_1\left(\eta_1 + \dfrac{D_2}{D_1}\eta_2\,\varepsilon_3\right) - t_3(n-i)L}{L^2}\right],$$

$$\omega_{i'}^{00} = -\frac{\rho d_1}{k_1}\left[\frac{\varepsilon_1}{a_1 + \beta_1 + \dfrac{D_3\,d_1}{D_1\,d_2}\varepsilon_1} - \frac{u_1\,M_1\,\varepsilon_3}{d_1\,t_3^2}\cdot\frac{d_1\left(\eta_1 + \dfrac{D_2}{D_1}\eta_2\,\varepsilon_3\right) - t_3(n-i)L}{L^2}\right],$$

$$\qquad\qquad\qquad \text{... (3.120)}$$

$$\psi_i^{00} = -\frac{\rho d_1}{k_1}\left[\frac{t_1}{a_1 + \beta_1 + \dfrac{D_3\,d_1}{D_1\,d_2}\varepsilon_1} - \frac{u_1\,M_1}{d_1\,t_3^2}\cdot\frac{d_1\left(\eta_1 + \dfrac{D_2}{D_1}\eta_2\,\varepsilon_3\right) - t_3(n-i)L}{L^2}\right].$$

By combining the general with the partial solution, we derive the following equation:

$$\omega_i = C_1\lambda_1^i + C_2\lambda_1^{-i} + C_3\lambda_2^i + C_4\lambda_2^{-i} + C_5\lambda_3^i + C_6\lambda_3^{-i} -$$

$$\frac{\rho d_1}{k_1}\left[\frac{1}{a_1 + \beta_1 + \dfrac{D_3\,d_1}{D_1\,d_2}\varepsilon_1} - \frac{u_1\,M_1}{d_1\,t_3^2}\cdot\frac{d_1\left(\eta_1 + \dfrac{D_2}{D_1}\eta_2\,\varepsilon_3\right) - t_3(n-i)L}{L^2}\right],$$

$$\omega_{i'} = \varepsilon_1\left(C_1\lambda_1^i + C_2\lambda_1^{-i}\right) + \varepsilon_2\left(C_3\lambda_2^i + C_4\lambda_2^{-i}\right) + \varepsilon_3\left(C_5\lambda_3^i + C_6\lambda_3^{-i}\right) - \quad\text{... (3.121)}$$

$$-\frac{\rho d_1}{k_1}\left[\frac{\varepsilon_1}{a_1 + \beta_1 + \dfrac{D_3\,d_1}{D_1\,d_2}\varepsilon_1} - \frac{u_1\,M_1\,\varepsilon_3}{d_1\,t_3^2}\cdot\frac{d_1\left(\eta_1 + \dfrac{D_2}{D_1}\eta_2\,\varepsilon_3\right) - t_3(n-i)L}{L^2}\right].$$

$$\psi_i = t_1\left(C_1\lambda_1^i + C_2\lambda_1^{-1}\right) + t_2\left(C_3\lambda_2^i + C_4\lambda_2^{-i}\right) + t_3\left(C_5\lambda_3^i + C_6\lambda_3^{-i}\right) -$$

$$-\frac{\rho d_1}{k_1}\left[\frac{t_1}{a_1 + \beta_1 + \dfrac{D_3\,d_1}{D_1\,d_2}\varepsilon_1} - \frac{u_1 M_1}{d_1 t_3}\cdot\frac{d_1\left(\eta_1 + \dfrac{D_2}{D_1}\eta_2\varepsilon_3\right) - t_3(n-i)L}{L^3}\right].$$

Let us consider the following limiting conditions:

$$\omega_0 = 0, \quad \omega_{0'} = 0, \quad \psi_0 = 0, \qquad\qquad \text{... (3.122)}$$

and also the equilibrium equation for upper nodes of the strip of the frame and all the upper layers (Figs. 3.48, 3.49 and 3.50):

$$\sum M_{(i=n)} = 0, \quad \sum M_{(i'=n')} = 0, \quad \sum y_{(n)} = 0.$$

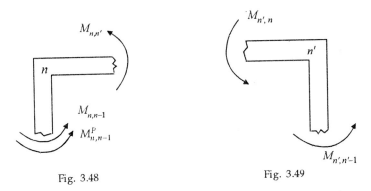

Fig. 3.48   Fig. 3.49

The three latter equations in the developed form are as follows:

$$\left(\alpha_1 + 4\frac{D_n}{D_1}\frac{d_1}{d_2}\right)\omega_n + \beta_1\,\omega_{n-1} - \frac{\gamma_1}{d_1}\psi_n + \frac{\eta_1}{d_1}\psi_{n-1} + 2\frac{D_n}{D_1}\frac{d_1}{d_2}\omega_{n'} = -\frac{\rho d_1}{k_1}A_1,$$

$$\left(\alpha_2 + 4\frac{D_n}{D_2}\frac{d_1}{d_2}\right)\omega_{n'} + \beta_1\,\omega_{n'-1} - \frac{\gamma_2}{d_1}\psi_n + \frac{\eta_2}{d_1}\psi_{n-1} + 2\frac{D_n}{D_1}\frac{d_1}{d_2}\omega_n = 0,$$

$$\gamma_1\,\omega_n + \eta_1\,\omega_{n-1} + \frac{D_2}{D_1}\omega_{n'} + \frac{D_2}{D_1}\eta_2\,\omega_{n'-1} - \frac{1}{d_1}\left(\mu_1 + \frac{D_2}{D_1}\mu_2 + 2r\frac{d_1^3}{D_1}\right)\psi_n +$$

$$\frac{1}{d_1}\left(\zeta_1 + \frac{D_2}{D_1}\zeta_2\right)\psi_{n-1} = -\frac{\rho.u_1}{k_1}\cdot G_1. \qquad \ldots (3.123)$$

Using conditions (3.122), we have:

$$C_1 + C_2 + C_3 + C_4 + C_5 + C_6 - \frac{\rho d_1}{k_1} \times$$

$$\left[\cfrac{1}{a_1 + \beta_1 + \cfrac{D_3}{D_1}\cfrac{d_1}{d_2}\varepsilon_1} - \frac{u_1 M_1}{d_1 t_3^2}\cdot\cfrac{d_1\left(\eta_1 + \cfrac{D_2}{D_1}\cfrac{d_1}{d_2}\eta_2\,\varepsilon_3\right) - t_3\,Ln}{L^2}\right] = 0,$$

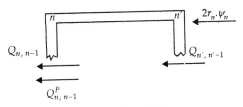

Fig. 3.50

138

$$\varepsilon_1(C_1 + C_2) + \varepsilon_2(C_3 + C_4) + \varepsilon_3(C_5 + C_6) - \frac{\rho d_1}{k_1} \times$$

$$\left[\frac{\varepsilon_1}{a_1 + \beta_1 + \dfrac{D_3}{D_1}\dfrac{d_1}{d_2}\varepsilon_1} - \frac{u_1 M_1 \varepsilon_3}{d_1 t_3^2} \cdot \frac{d_1\left(\eta_1 + \dfrac{D_2}{D_1}\eta_2\varepsilon_3\right) - t_3 Ln}{L^2} \right] = 0,$$

$$t_1(C_1 + C_2) + t_2(C_3 + C_4) + t_3(C_5 + C_6) - \frac{\rho d_1}{k_1} \times$$

$$\left[\frac{t_1}{a_1 + \beta_1 + \dfrac{D_3}{D_1}\dfrac{d_1}{d_2}\varepsilon_1} - \frac{u_1 M_1}{d_1 t_3^2} \cdot \frac{d_1\left(\eta_1 + \dfrac{D_2}{D_1}\dfrac{d_1}{d_2}\eta_2\varepsilon_3\right) - t_3 Ln}{L^2} \right] = 0.$$

Resolving this system of equations we find that:

$$C_5 + C_6 = \frac{\rho d_1}{k_1} \cdot \frac{u_1 M_1}{d_1 t_3^2} \cdot \frac{d_1\left(\eta_1 + \dfrac{D_2}{D_1}\eta_2\varepsilon_3\right) - t_3 Ln}{L^2},$$

$$C_3 + C_4 = 0, \quad C_1 + C_2 = \frac{\rho d_1}{k_1} \frac{1}{a_1 + \beta_1 + \dfrac{D_3}{D_1}\dfrac{d_1}{d_2}\varepsilon_1}. \qquad \text{... (3.124)}$$

By subjecting solution (3.121) to conditions (3.123), we derive:

$$a_1(1)C_1\lambda_1'' + b_1(1)C_2\lambda_1^{-''} + a_1(2)C_3\lambda_2^{-''} + b_1(2)C_4\lambda_2^{-''} + a_1(3)C_5\lambda_3'' + b_1(3)C_6\lambda_3^{-''} =$$

$$\frac{\rho d_1}{k_1}\left[\frac{f_1(1)}{a_1 + \beta_1 + \dfrac{D_3}{D_1}\dfrac{d_1}{d_2}\varepsilon_1} - \frac{u_1 M_1}{d_1 t_3^2} \cdot \frac{d_1\left(\eta_1 + \dfrac{D_2}{D_1}\eta_2\varepsilon_3\right)\cdot f_1(3) - t_3 L.\left(\beta_1 + \dfrac{t_3}{d_1}\eta_1\right)}{L^2} - A_1 \right],$$

$$a_2(1)C_1\lambda_1'' + b_2(1)C_2\,\lambda_1^{-''} + a_2(2)C_3\,\lambda_2'' + b_2(2)C_4\,\lambda_2^{-''} + a_2(3)C_5\lambda_3'' + b_1(3)C_6\lambda_3^{-''} =$$

$$\frac{\rho d_1}{k_1}\left[\frac{f_2(1)}{a_1+\beta_1+\dfrac{D_3}{D_1}\dfrac{d_1}{d_2}\varepsilon_1}-\frac{u_1 M_1}{d_1 t_3^2}\cdot\frac{d_1\left(\eta_1+\dfrac{D_2}{D_1}\eta_2\,\varepsilon_3\right)\cdot f_2(3)-t_3\,L\cdot\left(\beta_2\,\varepsilon_3+\dfrac{t_3}{d_1}\eta_2\right)}{L^2}\right],$$

$$(3.125)$$

$$a_3(1)C_1\,\lambda_1^n+b_3(1)C_2\,\lambda_1^{-n}+a_3(2)C_3\,\lambda_2^n+b_3(2)C_4\,\lambda_2^{-n}+a_3(3)C_5\,\lambda_3^n+b_3(3)C_6\,\lambda_3^{-n}=$$

$$=\frac{\rho d_1}{k_1}\left[\frac{f_3(1)}{a_1+\beta_1+\dfrac{D_3}{D_1}\dfrac{d_1}{d_2}\varepsilon_1}-\frac{u_1 M_1}{d_1 t_3^2}\times\right.$$

$$\left.\frac{d_1\left(\eta_1+\dfrac{D_2}{D_1}\eta_2\,\varepsilon_3\right)\cdot f_3(3)-t_3\,L\cdot\left[\left(\eta_1+\dfrac{D_2}{D_1}\eta_2\,\varepsilon_3\right)+\dfrac{t_3}{d_1}\left(\zeta_1+\dfrac{D_2}{D_1}\zeta_2\right)\right]}{L^2}-\frac{u_1}{d_1}\cdot G_1\right],$$

where

$$a_1(j)=\alpha_1-\frac{t_j}{d_1}\gamma_1+\left(\beta_1+\frac{t_j}{d_1}\eta_1\right)\frac{1}{\lambda_j}+2\frac{D_n}{D_1}\cdot\frac{d_1}{d_2}(2+\varepsilon_j),$$

$$a_2(j)=\alpha_2\varepsilon_j-\frac{t_j}{d_1}\gamma_2+\left(\varepsilon_j\beta_2+\frac{t_j}{d_1}\eta_2\right)\frac{1}{\lambda_j}+2\frac{D_n}{D_1}\cdot\frac{d_1}{d_2}(2\varepsilon_j+1)$$

$$a_3(j)=\gamma_1+\frac{D_2}{D_1}\gamma_2\varepsilon_j+\left[\eta_1+\frac{D_2}{D_1}\eta_2\,\varepsilon_j+\frac{t_j}{d_1}\left(\zeta_1+\frac{D_2}{D_1}\zeta_2\right)\right]\times$$

$$\frac{1}{\lambda_j}-\frac{t_j}{d_1}\left(\mu_1+\frac{D_2}{D_1}\mu_2+2r\frac{d_1^3}{D_1}\right),$$

$$b_1(j)=\alpha_1-\frac{t_j}{d_1}\gamma_1+\left(\beta_1+\frac{t_j}{d_1}\eta_1\right)\lambda_j+2\frac{D_n}{D_2}\cdot\frac{d_1}{d_2}(2+\varepsilon_j),$$

$$b_2(j)=\alpha_1\varepsilon_1-\frac{t_j}{d_1}\gamma_2+\left(\varepsilon_j\beta_2+\frac{t_j}{d_1}\eta_2\right)\lambda_j+2\frac{D_n}{D_2}\cdot\frac{d_1}{d_2}(2\varepsilon_j+1)$$

$$b_3(j)=\gamma_1+\frac{D_2}{D_1}\gamma_2\varepsilon_j+\left[\eta_1+\frac{D_2}{D_1}\eta_2\,\varepsilon_j+\frac{t_j}{d_1}\left(\zeta_1+\frac{D_2}{D_1}\zeta_2\right)\right]\times$$

$$\lambda_j - \frac{t_j}{d_1}\left(\mu_1 + \frac{D_2}{D_1}\mu_2 + 2r\frac{d_1^3}{D_1}\right).$$

$j = 1, 2, 3;$

$$f_1(j) = \alpha_1 + \beta_1 + \frac{t_j}{d_1}(\eta_1 - \gamma_1) + 2\frac{D_n}{D_1}\cdot\frac{d_1}{d_2}(2 + \varepsilon_j), \qquad \ldots (3.126)$$

$$f_2(j) = (\alpha_2 + \beta_2)\varepsilon_j + \frac{t_j}{d_1}(\eta_2 - \gamma_2) + 2\frac{D_n}{D_2}\cdot\frac{d_1}{d_2}(2\varepsilon_j + 1)$$

$$f_3(j) = \gamma_1 + \eta_1 + \frac{D_2}{D_1}(\gamma_2 - \eta_2)\varepsilon_j - \frac{t_j}{d_1}\left(\mu_1 + \frac{D_2}{D_1}\mu_2 + 2r_n\frac{d_1^3}{D_1}\right) + \frac{t_j}{d_1}\left(\zeta_1 + \frac{D_2}{D_1}\zeta_2\right),$$

$j = 1, 3.$

Considering eqns. (3.124) eqns. (3.125) assume the form:

$$C_1\left[a_1(1)\lambda_1^n - b_1(1)\lambda_1^{-n}\right] + C_3\left[a_1(2)\lambda_2^n - b_1(2)\lambda_2^{-n}\right] + C_5\left[a_1(3)\lambda_3^n - b_1(3)\lambda_3^{-n}\right] =$$

$$= \frac{\rho d_1}{k_1}\left\{\frac{f_1(1) - b_1(1)\lambda_1^{-n}}{a_1 + \beta_1 + \frac{D_3}{D_1}\frac{d_1}{d_2}\varepsilon_1} - \frac{u_1 M_1}{d_1 t_3^2}\times\right.$$

$$\left.\frac{d_1\left(\eta_1 + \frac{D_2}{D_1}\eta_2\varepsilon_3\right)\left[f_1(3) - b_1(3)\lambda_3^{-n}\right] - t_3\cdot L\cdot\left[\beta_1 + \frac{t_3}{d_1}\eta_1 - b_1(3)\lambda_3^{-n}\cdot n\right]}{L^2} - A_1\right\},$$

$$C_1\left[a_2(1)\lambda_1^n - b_2(1)\lambda_1^{-n}\right] + C_3\left[a_2(2)\lambda_2^n - b_2(2)\lambda_2^{-n}\right] + C_5\left[a_2(3)\lambda_3^n - b_2(3)\lambda_3^{-n}\right] =$$

$$\frac{\rho d_1}{k_1}\left\{\frac{f_2(1) - b_2(1)\lambda_1^{-n}}{a_1 + \beta_1 + \frac{D_3}{D_1}\frac{d_1}{d_2}\varepsilon_1} - \frac{u_1 M_1}{d_1 t_3^2}\times\right. \qquad \ldots (3.127)$$

$$\left.\frac{d_1\left(\eta_1 + \frac{D_2}{D_1}\eta_2\varepsilon_3\right)\left[f_2(3) - b_2(3)\lambda_3^{-n}\right] - t_3\cdot L\cdot\left[\beta_2\varepsilon_3 + \frac{t_3}{d_1}\eta_2 - b_2(3)\lambda_3^{-n}\cdot n\right]}{L^2}\right\},$$

$$C_1\left[a_3(1)\lambda_1^n - b_3(1)\lambda_1^{-n}\right] + C_3\left[a_3(2)\lambda_2^n - b_3(2)\lambda_2^{-n}\right] + C_5\left[a_3(3)\lambda_3^n - b_3(3)\lambda_3^{-n}\right] =$$

$$\frac{\rho d_1}{k_1} \left\{ \frac{f_3(1) - b_3(1)\lambda_1^{-n}}{a_1 + \beta_1 + \dfrac{D_3}{D_1}\dfrac{d_1}{d_2}\varepsilon_1} - \frac{u_1 M_1}{d_1 t_3^2} \times \right.$$

$$\times d_1 \left(\eta_1 + \frac{D_2}{D_1^2}\eta_2\varepsilon_3 \right) \left[f_3(3) - b_3(3)\lambda_3^{-n} \right] - t_3 \cdot L \times$$

$$\left. \times \frac{\left[\eta_1 + \dfrac{D_2}{D_1^2}\eta_2\varepsilon_3 + \dfrac{t_3}{d_1}\left(\zeta_1 + \dfrac{D_2}{D_1}\zeta_2 \right) - b_1(3)\lambda_3^{-n}\cdot n \right]}{L^2} - \frac{u_1}{d_1}\cdot G_1 \right\}$$

Constants C_1, C_3 and C_5 are determined by solving the system of equations (3.127). Later, from eqn. (3.124), we find C_2, C_4 and C_6:

$$C_2 = \frac{\rho d_1}{k_1} \cdot \frac{1}{a_1 + \beta_1 + \dfrac{D_3}{D_1}\dfrac{d_1}{d_2}\varepsilon_1} - C_1, \quad C_4 = -C_3,$$

$$C_6 = \frac{\rho}{k_1}\cdot\frac{u_1 M_1}{t_3^2} \frac{d_1\left(\eta_1 + \dfrac{D_2}{D_1}\eta_2\,\varepsilon_3 \right) - t_3\cdot L\cdot n}{L^2} - C_5. \qquad \text{... (3.128)}$$

By substituting the values of constants in eqn. (3.121), we can find the angular and linear radial displacements of all nodes on the strip of the frame.

To determine the forces and displacements of the sections of branches located between the nodes, equations of the method of initial parameters (3.102) can be used. We then have for the panel bound by nodes $i-1$ and i:

$$M = M_0 \tilde{A} + \tilde{V}_0 B + U_0 \tilde{C} + W_0 \tilde{D}, \qquad \text{... (3.129)}$$

$$U = -4 M_0 \tilde{C} - 4 V_0 \tilde{D} + U_0 \tilde{A} + W_0 \tilde{B}$$

in which
—for elements of branch No. 1:

$$U_0 = \frac{\rho d_1 (n - i + s/d_1) + k_1 \cdot y_1 \cdot l}{u_1^2},$$

$$M_0 = \frac{D_1}{d_1}\left(\alpha_1 \omega_i + \beta_1 \omega_{i-1} - \gamma_1 \frac{\psi_i}{d_1} + \eta_1 \frac{\psi_{i-1}}{d_1} \right),$$

$$V_0 = \frac{D_1}{u_1 d_1^2}\left(\gamma_1\omega_i + \eta_1\omega_{i-1} - \mu_1\frac{\psi_i}{d_1} + \zeta_1\frac{\psi_{i-1}}{d_1}\right), \quad U_0 = \frac{\rho d_1(n-i) + k_1\psi_i \cdot l}{u_1^2},$$

$$W_0 = \frac{\rho + k_1\omega_i \cdot l}{u_1^3}, \qquad\qquad\qquad\qquad\qquad \text{... (3.130)}$$

—for elements of branch No. 2:

$$U = \frac{k_2 \cdot y_2 \cdot l}{u_2^2} \quad M_0 = \frac{D_2}{d_1}\left(\alpha_2\omega_{i'} + \beta_2\omega_{i'-1} - \gamma_2\frac{\psi_i}{d_1} + \eta_2\frac{\psi_{i-1}}{d_1}\right),$$

$$V_0 = \frac{D_2}{u_2 d_1^2}\left(\gamma_2\omega_{i'} + \eta_2\omega_{i'-1} - \mu_2\frac{\psi_i}{d_1} + \zeta_2\frac{\psi_{i-1}}{d_1}\right), \qquad \text{... (3.131)}$$

$$U_0 = \frac{k_2 \cdot \psi_i \cdot l}{u_2^2}, \quad W_0 = \frac{k_2 \cdot \omega_i \cdot l}{u_2^3}.$$

Taking into account eqn. (3.130), we derive the following equations for the bending moment and deflection of an arbitrary section of branch No. 1 between nodes $i-1$ and i:

$$M_{(\bar{s}_1)} = \frac{D_1}{d_1}\left(\alpha_1\omega_i + \beta_1\omega_{i-1} - \gamma_1\frac{\psi_i}{d_1} + \eta_1\frac{\psi_{i-1}}{d_1}\right)\cdot A(\bar{s}_1) -$$

$$\frac{D_1}{u_1 d_1^2}\left(\gamma_1\omega_i + \eta_1\omega_{i-1} - \mu_1\frac{\psi_i}{d_1} + \zeta_1\frac{\psi_{i-1}}{d_1}\right)\cdot B(\bar{s}_1) +$$

$$\frac{\rho d_1(n-i) + k_1\psi_i}{u_1^2}\cdot C(\bar{s}_1) + \frac{\rho + k_1\omega_i}{u_1^3}\cdot D(\bar{s}_1), \qquad \text{... (3.132)}$$

$$y(\bar{s}_1) = \frac{u_1^2}{k_1}\left[-\frac{4D_1}{d_1}\left(\alpha_1\omega_i + \beta_1\omega_{i-1} - \gamma_1\frac{\psi_i}{d_1} + \eta_1\frac{\psi_{i-1}}{d_1}\right)\cdot C(\bar{s}_1) +\right.$$

$$\frac{4D_1}{u_1 d_1^2}\left(\gamma_1\omega_i + \eta_1\omega_{i-1} - \mu_1\frac{\psi_i}{d_1} + \zeta_1\frac{\psi_{i-1}}{d_1}\right)\cdot D(\bar{s}_1) +$$

$$\left.\frac{\rho d_1(n-i) + k_1\psi_i}{u_1^2}\cdot A(\bar{s}_1) + \frac{\rho + k_1\omega_i}{u_1^3}\cdot B(\bar{s}_1)\right] - \frac{\rho d_1}{k_1}\left(n - i + \frac{s}{d_1}\right).$$

Correspondingly, in the section of branch No. 2 between nodes $i' = 1$ and i':

$$M(\bar{s}_2) = \frac{D_2}{d_1}\left(\alpha_2\omega_{i'} + \beta_2\omega_{i'-1} - \gamma_2\frac{\psi_i}{d_1} + \eta_2\frac{\psi_{i-1}}{d_1}\right)\cdot A(\bar{s}_2) -$$

$$\frac{D_2}{u_2 d_1^2}\left(\gamma_2\,\omega_{i'} + \eta_2\,\omega_{i'-1} - \mu_2\frac{\psi_i}{d_1} + \zeta_2\frac{\psi_{i-1}}{d_1}\right)\cdot B(\bar{s}_2) +$$

$$\frac{k_2\,\psi_i}{u_2^2}\cdot C(\bar{s}_2) + \frac{k_2\,\omega_{i'}}{u_2^3}\cdot D(\bar{s}_2), \qquad\qquad \dots (3.133)$$

$$y(\bar{s}_2) = \frac{u_2^2}{k_2}\left[-\frac{4D_2}{d_1}\left(\alpha_2\,\omega_{i'} + \beta_2\,\omega_{i'-1} - \gamma_2\frac{\psi_i}{d_1} + \eta_2\frac{\psi_{i-1}}{d_1}\right)\cdot C(\bar{s}_2) + \right.$$

$$\left[\frac{4D_2}{u_2 d_1^2}\left(\gamma_2\,\omega_{i'} + \eta_2\,\omega_{i'-1} - \mu_2\frac{\psi_i}{d_1} + \zeta_2\frac{\psi_{i-1}}{d_1}\right)\cdot D(\bar{s}_2) + \right.$$

$$\left.\frac{k_2\,\psi_i}{u_2^2}\cdot A(\bar{s}_2) + \frac{k_2\,\omega_{i'}}{u_2^3}\cdot B(\bar{s}_2)\right].$$

The following notations were used in eqns. (3.132) and (3.133):

$$A(\bar{s}_1) = \cos\bar{s}_1\,ch\,\bar{s}_1 \qquad\qquad A(\bar{s}_2) = \cos\bar{s}_2\cdot ch\,\bar{s}_2$$

$$B(\bar{s}_1) = \frac{1}{2}\left(\sin\bar{s}_1\cdot ch\,\bar{s}_1 + \cos\bar{s}_1\,ch\,\bar{s}_1\right), \qquad B(\bar{s}_2) = \frac{1}{2}\left(\sin\bar{s}_2\cdot ch\,\bar{s}_2 + \cos\bar{s}_2\,sh\,\bar{s}_2\right),$$

$$C(\bar{s}_1) = \frac{1}{2}\sin\bar{s}_1\cdot sh\,\bar{s}_1, \qquad\qquad C(\bar{s}_2) = \frac{1}{2}\sin\bar{s}_2\cdot sh\,\bar{s}_2, \qquad \dots (3.134)$$

$$D(\bar{s}_1) = \frac{1}{4}\left(\sin\bar{s}_1\cdot ch\,\bar{s}_1 - \cos\bar{s}_1\cdot sh\,\bar{s}_1\right), \qquad D(\bar{s}_1) = \frac{1}{4}\left(\sin\bar{s}_2\cdot ch\,\bar{s}_2 - \cos\bar{s}_2\cdot sh\,\bar{s}_2\right),$$

$$\bar{s}_1 = u_1\cdot s, \qquad\qquad \bar{s}_2 = u_2\cdot s.$$

By knowing the radial displacements of arbitrary sections of branches of the strip of the frame $y(\bar{s}_1)$ and $y(\bar{s}_2)$, the current values of forces in the peripheral direction can be determined (Fig. 3.40):

$$T(\bar{s}_1) = \frac{E\delta_1}{R_2}\cdot y(\bar{s}_1), \quad T(\bar{s}_2) = \frac{E\delta_2}{R_2}\cdot y(\bar{s}_2). \qquad \dots (3.135)$$

Thus, the derived eqns. (3.132), (3.133) and (3.135) help to accurately investigate the stressed and deformed states of any structure comprising two coaxial cylindrical shells of different thicknesses joined by frames and subjected to the action of axisymmetrical hydrostatic loading.

3.6.2 A Case of Walls of Identical Thickness

When the gap between the co-axial shells is comparatively small $(d_2/2R \le 0.025)$, the difference in the values of the inertial moments of the chords of a single strip does not exceed 5%, i.e., $(2R - d_2)/(2R + d_2) \ge 0.95$.

The coefficients of the bases of branches k_1 and k_2 will also differ by not more than 5%. In this case, an elementary strip of the frame can be regarded with a fair degree of accuracy as symmetrical with branches having identical parameters:

$\delta_1 = \delta_2 = \delta$, $D_1 = D_2 = D$, $k_1 = k_2 = k$. Then, $\alpha_1 = \alpha_2 = \alpha$, $\beta_1 = \beta_2 = \beta$, $\gamma_1 = \gamma_2 = \gamma$, $\eta_1 = \eta_2 = \eta$, $\mu_1 = \mu_2 = \mu$, $\zeta_1 = \zeta_2 = \zeta$, $\alpha_1 = \alpha_2 = \alpha$.

Considering these equations, it follows from eqn. (3.116)

$$\varepsilon = \frac{\eta^2 \left(1 \cdot \lambda^2\right)^2 - 2\left[2a\left(\lambda - \lambda^3\right) + \beta\left(1 - \lambda^4\right)\right]\zeta + 2\left[2a\lambda^2 + \beta\left(\lambda + \lambda^3\right)\right]\left(2\mu + r\dfrac{d_1^4}{D}\right)}{\eta^2\left(1 - \lambda^2\right)^2 + 4\dfrac{D_3}{D}\cdot\dfrac{d_1}{d_2}\left(2\mu + r\dfrac{d_1^4}{D}\right)\lambda^2 - 4\dfrac{D_3}{D}\cdot\dfrac{d_1}{d_2}\zeta\left(\lambda - \lambda^3\right)},$$

$$t = -d_1 \cdot \frac{2\eta\left(a - \dfrac{D_3}{D}\cdot\dfrac{d_1}{d_2}\right)\left(\lambda - \lambda^3\right) + \beta\eta\left(1 - \lambda^4\right)}{\eta^2\left(1 - \lambda^2\right)^2 + 4\dfrac{D_3}{D}\cdot\dfrac{d_1}{d_2}\left(2\mu + r\dfrac{d_1^4}{D}\right)\lambda^2 - 4\dfrac{D_3}{D}\cdot\dfrac{d_1}{d_2}\left(\lambda - \lambda^3\right)\zeta}. \quad \ldots (3.136)$$

The coefficients of characteristic eqn. (3.109) will be:

$$A_0 = \frac{4a\beta\zeta - 2a\eta^2 - \beta^2\left(2\mu + r\dfrac{d_1^3}{D}\right) + 2\dfrac{D_3}{D}\cdot\dfrac{d_1}{d_2}\eta^2}{\beta\left(\beta\zeta - \eta^2\right)},$$

$$B_0 = \frac{\left(4a^2 + 3\beta^2 - 4\dfrac{D_3^2}{D^2}\cdot\dfrac{d_1^2}{d_2^2}\right)\zeta - 4a\beta\left(2\mu + r\dfrac{d_1^3}{D}\right) + \beta\eta^2}{\beta\left(\beta\zeta - \eta^2\right)}, \quad \ldots (3.137)$$

$$C_0 = \frac{4a\beta\zeta - \left(2a^2 + \beta^2 - 2\dfrac{D_3^2}{D^2}\cdot\dfrac{d_1^2}{d_2^2}\right)\left(2\mu + r\dfrac{d_1^3}{D}\right) + 2\left(a - \dfrac{D_3}{D}\cdot\dfrac{d_1}{d_2}\right)\eta^2}{\beta\left(\beta\zeta - \eta^2\right)},$$

$$a = \alpha + 2\frac{D_3}{D}\cdot\frac{d_1}{d_2}.$$

The solution to the system of eqns. (3.107) will be:

$$\omega_i = C_1\lambda_1^i + C_2\lambda_1^{-i} + C_3\lambda_2^i + C_4\lambda_2^{-i} + C_5\lambda_3^i + C_6\lambda_3^{-i} -$$

$$\frac{\rho d_1}{k}\left[\frac{1}{a+\beta+\dfrac{D_3}{D}\cdot\dfrac{d_1}{d_2}\delta_1}-\frac{u\cdot M}{d_1 t_3^2}\cdot\frac{d_1\,\eta(1+\varepsilon_3)-t_3\cdot L\cdot(n-i)}{L^2}\right],$$

$$\omega_{i'}=\varepsilon_1\left(C_1\,\lambda_1^i+C_2\,\lambda_1^{-i}\right)+\varepsilon_2\left(C_3\,\lambda_2^i+C_4\,\lambda_2^{-i}\right)+\varepsilon_3\left(C_5\,\lambda_3^i+C_6\,\lambda_3^{-i}\right)-$$

$$\frac{\rho d_1}{k}\left[\frac{\varepsilon_1}{a+\beta+\dfrac{D_3}{D}\cdot\dfrac{d_1}{d_2}\varepsilon_1}-\frac{u\cdot M\cdot\varepsilon_3}{d_1 t_3^2}\cdot\frac{d_1\,\eta(1+\varepsilon_3)-t_3\cdot L\cdot(n-i)}{L^2}\right],$$

$$\psi_i=t_1\left(C_1\,\lambda_1^i+C_2\,\lambda_1^{-i}\right)+t_2\left(C_3\,\lambda_2^i+C_4\,\lambda_2^{-i}\right)+t_3\left(C_5\,\lambda_3^i+C_6\,\lambda_3^{-i}\right)-$$

$$\frac{\rho d_1}{k}\left[\frac{t_1}{a+\beta+\dfrac{D_3}{D}\cdot\dfrac{d_1}{d_2}\varepsilon_1}-\frac{u\cdot M}{d_1 t_3^2}\cdot\frac{d_1\,\eta(1+\varepsilon_3)-t_3\cdot L\cdot(n-i)}{L^2}\right], \qquad \ldots (3.138)$$

$$L=2(\zeta-\mu)-rd_1^3/D.$$

The limiting conditions in a developed form assume the following form:

$$C_1\left[a_1(1)\,\lambda^n-b_1(1)\lambda_1^{-n}\right]+C_3\left[a_1(2)\,\lambda_2^n-b_1(2)\lambda_2^{-n}\right]+C_5\left[a_1(3)\lambda_3^n-b_1(3)\,\lambda_3^{-n}\right]=$$

$$\frac{\rho d_1}{k}\left\{\frac{f_1(1)-b_1(1)\,\lambda_1^{-n}}{a+\beta+\dfrac{D_3}{D}\cdot\dfrac{d_1}{d_2}\varepsilon_1}-\frac{u\cdot M}{d_1 t_3^2}\times\right.$$

$$\left.\frac{d_1\eta(1+\varepsilon_3)\left[f_1(3)-b_1(3)\,\lambda_3^{-n}\right]-t_3\cdot L\cdot\left[\beta+\dfrac{t_3}{d_1}\eta-n\cdot b_2(3)\cdot\lambda_3^{-n}\right]}{L^2}-A\right\},$$

$$C_1\left[a_2(1)\,\lambda_1^n-b_2(1)\,\lambda_1^{-n}\right]+C_3\left[a_2(2)\,\lambda_2^n-b_2(2)\,\lambda_2^{-n}\right]+C_5\left[a_2(3)\,\lambda_3^n-b_2(3)\,\lambda_3^{-n}\right]=$$

$$=\frac{\rho d_1}{k}\left\{\frac{f_2(1)-b_2(1)\lambda_1^{-n}}{\alpha+\beta+\dfrac{D_3}{D}\cdot\dfrac{d_1}{d_2}\varepsilon_1}-\frac{u\cdot M}{d_1 t_3^3}\times\right.$$

$$\left.\frac{d_1\eta(1 + \varepsilon_3)\left[f_2(3) - b_2(3)\,\lambda_3^{-n}\right] - t_3\cdot L\cdot\left[\beta\varepsilon_3 + \dfrac{t_3}{d_1}\eta - n\cdot b_2\,(3)\cdot\lambda_3^{-n}\right]}{L^2}\right\},$$

$$C_1\left[a_3(1)\,\lambda_1^{n} - b_3(1)\,\lambda_1^{-n}\right] + C_3\left[a_3(2)\,\lambda_2^{n} - b_3(2)\,\lambda_2^{-n}\right] + C_5\left[a_3(3)\,\lambda_3^{n} - b_3(3)\,\lambda_3^{-n}\right] =$$

$$= \frac{\rho d_1}{k}\left\{\frac{f(3) - b_3(1)\,\lambda_1^{-n}}{\alpha + \beta + \dfrac{D_3}{D}\cdot\dfrac{d_1}{d_2}\,\varepsilon_1} - \frac{u\cdot M}{d_1\,t_3^2}\times\right.$$

$$\left.\frac{d_1\eta(1 + \varepsilon_3)\left[f_2(3) - b_3(3)\,\lambda_3^{-n}\right] - t_3\cdot L\cdot\left[\eta(1 + \varepsilon_3) + 2\dfrac{t_3}{d_1}\xi - n\cdot b_3\,(3)\lambda_3^{-n}\right]}{L^2} - u\cdot\Sigma\right\},$$

$$C_2 = \frac{\rho d_1}{k}\cdot\frac{1}{a + \beta + \dfrac{D_3}{D}\cdot\dfrac{d_1}{d_2}\,\varepsilon_1} - C_1, \qquad\qquad C_4 = -C_3,$$

$$C_6 = -\frac{\rho}{k}\cdot\frac{u\cdot M}{t_3^2}\cdot\frac{d_1\,\eta(1 + \varepsilon_3) - t_3\cdot n\cdot L}{L_2} - C_5. \qquad\qquad \text{... (3.139)}$$

The following equations are derived for bending moments, deflections and forces in arbitrary section of branches between nodes i–1 and i:

$$M_1(\bar{s}) = \frac{D}{d_1}\left(\alpha\cdot\omega_i + \beta\omega_{i-1} - \gamma\frac{\psi_i}{d_1} + \eta\frac{\psi_{i-1}}{d_1}\right)\cdot A(\bar{s}) - \frac{D}{ud_1^2}\times$$

$$\left(\gamma\omega_1 + \eta\omega_{i-1} - \mu\frac{\psi_i}{d_1} + \zeta\frac{\psi_{i-1}}{d_1}\right)\cdot B(\bar{s}) + \frac{\rho d_1(n - i) + k\cdot\psi_i}{u^2}\cdot C(\bar{s}) +$$

$$\frac{\rho + k\cdot\omega_i}{u^3}\cdot D(\bar{s}), \qquad\qquad \text{... (3.140)}$$

$$y_1(\bar{s}) = \frac{u^2}{k}\left[-\frac{4D}{d_1}\left(\alpha\cdot\omega_i + \beta\cdot\omega_{i-1} - \gamma\frac{\psi_i}{d_1} + \eta\frac{\psi_{i-1}}{d_1}\right)\cdot C(\bar{s}) +\right.$$

$$\frac{4D}{ud_1^2}\left(\gamma\cdot\omega_i + \eta\omega_{i-1} - \mu\frac{\psi_i}{d_1} + \zeta\frac{\psi_{i-1}}{d_1}\right)\cdot D(\bar{s}) + \frac{pd_1(n-i)+k\cdot\psi_i}{u^2}\cdot A(\bar{s}) +$$

$$\frac{p+k\cdot\omega_i}{u^3}\cdot B(\bar{s}) - \frac{pd_1}{k}(n-i+\bar{s}), \qquad\qquad T_1(\bar{s}) = k\cdot R_1\cdot y_1(\bar{s}),$$

$$M_2(\bar{s}) = \frac{D}{d_1}\left(\alpha\cdot\omega_{i'} + \beta\omega_{i'-1} - \gamma\frac{\psi_i}{d_1} + \eta\frac{\psi_{i-1}}{d_1}\right)\cdot A(\bar{s}) -$$

$$\frac{D}{ud_1^2}\left(\gamma\omega_i + \eta\omega_{i'-1} - \mu\frac{\psi_i}{d_1} + \zeta\frac{\psi_{i-1}}{d_1}\right)\cdot B(\bar{s}) + \frac{k}{u^2}\left[\psi_i\cdot C(\bar{s}) + \frac{1}{u}\cdot\omega_{i'}\cdot D(\bar{s})\right]$$

$$y_2(\bar{s}) = \frac{u^2}{k}\left[-\frac{4D}{d_1}\left(\alpha\cdot\omega_i + \beta\omega_{i-1} - \gamma\frac{\psi_i}{d_1} + \eta\frac{\psi_{i-1}}{d_1}\right)\cdot C(\bar{s}) +\right.$$

$$\frac{4D}{ud_1^2}\left(\gamma\omega_{i'} + \eta\omega_{i'-1} - \mu\frac{\psi_i}{d_1} + \zeta\frac{\psi_{i-1}}{d_1}\right)\cdot D(\bar{s}) + \frac{k}{u^2}\left[\psi_i A(\bar{s}) + \frac{1}{u}\cdot\omega_{i'}\cdot B(\bar{s})\right].$$

3.6.3 Analysis of a Cylindrical Compartment for External Pressure

Submarine systems with double walls are exposed to considerable external pressures. Let us examine a compartment of such a system with a cylindrical form represented by two coaxial shells joined by circular ribs or frames (Fig. 3.51).

Under conditions of axisymmetrical shell deformations, all the strips isolated along the generatrix of the cylinder will function under identical conditions like a frame with many panels whose branches are joined with elastic foundations and nodes with elastic supports (Fig. 3.52).

The equations for moments due to loads and transverse forces are derived as a partial case from equations shown in Section 3.5.1 at:

$$q_i = q = \text{const}: M^p_{i',i'-1} M^p_{i',i'+1} = \frac{qd_1^2}{4u_2^4}\cdot\gamma_2, \ Q^p_{i',i'-1} = Q^p_{i',i'+1} = -\frac{qd_1}{4u_2^3}\cdot\mu. \ \ldots \ (3.141)$$

Then, the equilibrium equations for nodes i', i and i-th strip will respectively be:

$$2\left(\alpha_2 + 2\frac{D_3}{D_2}\cdot\frac{d_1}{d_2}\right)\omega_{i'} + \beta_2(\omega_{i'-1}+\omega_{i'+1}) + \frac{\eta_2}{d_1}(\psi_{i-1}-\psi_{i+1}) + 2\frac{D_3}{D_2}\cdot\frac{d_1}{d_2}\omega_i = 0,$$

$$2\left(\alpha_1 + 2\frac{D_3}{D_1}\cdot\frac{d_1}{d_2}\right)\omega_i + \beta_1(\omega_{i-1}+\omega_{i+1}) + \frac{\eta_1}{d_1}(\psi_{i-1}-\psi_{i+1}) + 2\frac{D_3}{D_1}\cdot\frac{d_1}{d_2}\omega_{i'} = 0,$$

$$\eta_1(\omega_{i-1}+\omega_{i+1}) + \frac{D_2}{D_1}\eta_2(\omega_{i'-1}-\omega_{i'+1}) - \frac{2}{d_1}\left(\mu_1 + \frac{D_2}{D_1}\cdot\mu_2 + r\frac{d_1}{D}\right)\psi_i +$$

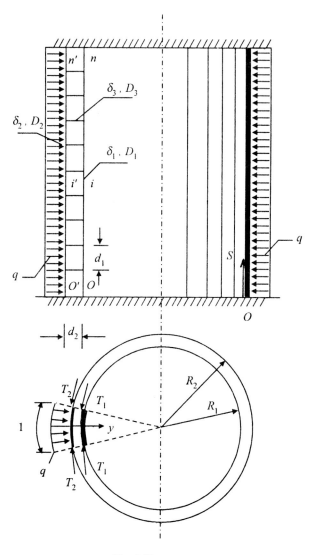

Fig. 3.51

$$\frac{1}{d_1}\left(\zeta_1 + \frac{D_2}{D_1}\zeta_2\right)(\psi_{i+1} + \psi_{i-1}) = -\frac{2u_2}{k_2 d_1} \cdot \frac{D_2}{D_1} \cdot \mu_2 \cdot q. \qquad \dots \text{(3.142)}$$

The general solution to the system of equations (3.142) will evidently coincide with the general solution of eqns. (3.107). However, we can find the partial solution in the form:

$$\omega_i^{00} = (A_1 + A_2 + A_3)q, \quad \omega_{i'}^{00} = (\varepsilon_1 A_1 + \varepsilon_2 A_2 + \varepsilon_3 A_3)q,$$

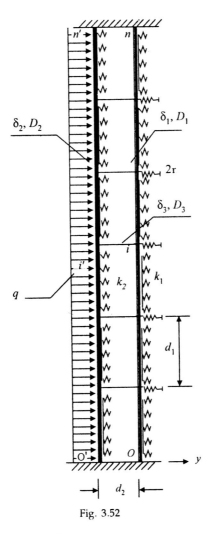

Fig. 3.52

$$\psi_i^{00} = (t_1 A_1 + t_2 A_2 + t_3 A_3) q.$$

After substituting in eqn. (3.142) and equating the coefficients for identical degrees of i, we find:

$$\omega_i^{00} = 0, \quad \omega_{i'}^{00} = 0, \quad \psi_i^{00} = -\frac{u_2}{k_2} \cdot \frac{D_2}{D_1} \cdot \frac{\mu_2}{L} \cdot q, \qquad \ldots \text{(3.143)}$$

$$L = \zeta_1 - \frac{D_2}{D_1} \cdot \zeta_2 - \mu_1 - \frac{D_2}{D_1} \mu_2 - r \frac{d_1^3}{D_1}.$$

The complete solution will then be:

$$\omega_1 = C_1 \lambda_1^i + C_2 \lambda_1^{-i} + C_3 \lambda_2^i + C_4 \lambda_2^{-i} + C_5 \lambda_3^i + C_6 \lambda_3^{-i},$$

$$\omega_{1'} = \varepsilon_1\left(C_1\,\lambda_1^i + C_2\lambda_1^{-i}\right) + \varepsilon_2\left(C_3\lambda_2^i + C_4\lambda_2^{-i}\right) + \varepsilon_3\left(C_5\,\lambda_3^i + C_6\,\lambda_3^{-i}\right), \qquad \dots \text{(3.144)}$$

$$\psi_1 = t_1\left(C_1\,\lambda_1^i + C_2\,\lambda_1^{-i}\right) + t_2\left(C_3\,\lambda_2^i + C_4\,\lambda_2^{-i}\right) + t_3\left(C_5\,\lambda_3^i + C_6\,\lambda_3^{-i}\right) -$$

$$\frac{u_2}{k_2}\cdot\frac{D_2}{D_1}\cdot\frac{\mu_2}{L}q,$$

in which the values of L are determined using eqn. (3.119) and ε_1, ε_2, ε_3, t_1, t_2 and t_3 from eqn. (3.106).

The conditions of fixing the ends of the strip of the frame in the diaphragms can be used as the limiting conditions (Fig. 3.52):

$$\omega_0 = 0, \qquad\qquad \omega_{0'} = 0, \qquad\qquad \psi_0 = 0, \qquad\qquad \dots \text{(3.145)}$$

$$\omega_n = 0, \qquad\qquad \omega_{n'} = 0, \qquad\qquad \psi_n = 0. \qquad\qquad \dots \text{(3.146)}$$

On satisfying eqn. (3.145), we have

$$C_1 + C_2 + C_3 + C_4 + C_5 + C_6 = 0,$$

$$\varepsilon_1(C_1 + C_2) + \varepsilon_2(C_3 + C_4) + \varepsilon_3(C_5 + C_6) = 0,$$

$$t_1(C_1 + C_2) + t_2(C_3 + C_4) + t_3(C_5 + C_6) = \frac{u_2}{k_2}\cdot\frac{D_2}{D_1}\cdot\frac{\mu_2}{L}\cdot q.$$

Hence,

$$C_1 + C_2 = \frac{u_2}{k_2}\cdot\frac{D_2}{D_1}\cdot\frac{\mu_2}{L}\cdot q\cdot\frac{\varepsilon_1 - \varepsilon_3}{t_1(\varepsilon_3 - \varepsilon_2) + t_2(\varepsilon_1 - \varepsilon_3) + t_3(\varepsilon_2 - \varepsilon_1)},$$

$$C_3 + C_4 = \frac{u_2}{k_2}\cdot\frac{D_2}{D_1}\cdot\frac{\mu_2}{L}\cdot q\cdot\frac{\varepsilon_1 - \varepsilon_3}{t_1(\varepsilon_3 - \varepsilon_2) + t_2(\varepsilon_1 - \varepsilon_3) + t_3(\varepsilon_2 - \varepsilon_1)},$$

$$C_5 + C_6 = \frac{u_2}{k_2}\cdot\frac{D_2}{D_1}\cdot\frac{\mu_2}{L}\cdot q\cdot\frac{\varepsilon_2 - \varepsilon_1}{t_1(\varepsilon_3 - \varepsilon_2) + t_2(\varepsilon_1 - \varepsilon_3) + t_3(\varepsilon_2 - \varepsilon_1)}. \qquad \dots \text{(3.147)}$$

Subjecting solution (3.144) to conditions (3.146), we find:

$$C_1\,\lambda_1^n + C_2\,\lambda_1^{-n} + C_3\,\lambda_2^n + C_4\,\lambda_2^{-n} + C_5\,\lambda_3^n + C_6\,\lambda_3^{-n} = 0,$$

$$\varepsilon_1\left(C_1\,\lambda_1^n + C_2\,\lambda_1^{-n}\right) + \varepsilon_2\left(C_3\,\lambda_2^n + C_4\,\lambda_2^{-n}\right) + \varepsilon_3\left(C_5\,\lambda_3^n + C_6\,\lambda_3^{-n}\right) = 0,$$

$$t_1\left(C_1\,\lambda_1^n + C_2\,\lambda_1^{-n}\right) + t_2\left(C_3\,\lambda_2^n + C_4\,\lambda_2^{-n}\right) + t_3\left(C_5\,\lambda_3^n + C_6\,\lambda_3^{-n}\right) =$$

$$\frac{u_2}{k_2}\cdot\frac{D_2}{D_1}\cdot\frac{\mu_2}{L}\cdot q$$

hence,

$$C_1\,\lambda_1^n + C_2\,\lambda_1^{-n} = \frac{u_2}{k_2}\cdot\frac{D_2}{D_1}\cdot\frac{\mu_2}{L}\cdot q\cdot\frac{\varepsilon_3 - \varepsilon_2}{t_1(\varepsilon_3 - \varepsilon_2) + t_2(\varepsilon_1 - \varepsilon_3) + t_3(\varepsilon_2 - \varepsilon_1)},$$

$$C_3 \lambda_2'' + C_4 \lambda_2^{-''} = \frac{u_2}{k_2} \cdot \frac{D_2}{D_1} \cdot \frac{\mu_2}{L} \cdot q \cdot \frac{\varepsilon_1 - \varepsilon_3}{t_1(\varepsilon_3 - \varepsilon_2) + t_2(\varepsilon_1 - \varepsilon_3) + t_3(\varepsilon_2 - \varepsilon_1)},$$

$$\dots (3.148)$$

$$C_5 \lambda_3'' + C_5 \lambda_3^{-''} = \frac{u_2}{k_2} \cdot \frac{D_2}{D_1} \cdot \frac{\mu_2}{L} \cdot q \cdot \frac{\varepsilon_2 - \varepsilon_1}{t_1(\varepsilon_3 - \varepsilon_2) + t_2(\varepsilon_1 - \varepsilon_3) + t_3(\varepsilon_2 - \varepsilon_1)}.$$

By deducting eqn. (3.148) from (3.147), we derive:

$$C_1 + C_2 = C_1 \lambda_1'' + C_2 \lambda_1^{-''},$$

$$C_3 + C_4 = C_3 \lambda_2'' + C_4 \lambda_2^{-''},$$

$$C_5 + C_6 = C_5 \lambda_3'' + C_6 \lambda_3^{-''},$$

it thus follows:

$$C_2 = C_1 \lambda_1'', \quad C_4 = C_3 \lambda_2'', \quad C_6 = C_5 \lambda_3''. \qquad \dots (3.149)$$

Later, from eqn. (3.147) or (3.148) and allowing for (3.149), we find

$$C_1 = \frac{u_2}{k_2} \cdot \frac{D_2}{D_1} \cdot \frac{\mu_2}{L} \cdot \frac{q}{\lambda_1'' + 1} \cdot \frac{\varepsilon_3 - \varepsilon_2}{t_1(\varepsilon_3 - \varepsilon_2) + t_2(\varepsilon_1 - \varepsilon_3) + t_3(\varepsilon_2 - \varepsilon_1)},$$

$$C_3 = \frac{u_2}{k_2} \cdot \frac{D_2}{D_1} \cdot \frac{\mu_2}{L} \cdot \frac{q}{\lambda_2'' + 1} \cdot \frac{\varepsilon_1 - \varepsilon_3}{t_1(\varepsilon_3 - \varepsilon_2) + t_2(\varepsilon_1 - \varepsilon_3) + t_3(\varepsilon_2 - \varepsilon_1)}, \qquad \dots (3.150)$$

$$C_4 = \frac{u_2}{k_2} \cdot \frac{D_2}{D_1} \cdot \frac{\mu_2}{L} \cdot \frac{q}{\lambda_3'' + 1} \cdot \frac{\varepsilon_2 - \varepsilon_1}{t_1(\varepsilon_3 - \varepsilon_2) + t_2(\varepsilon_1 - \varepsilon_3) + t_3(\varepsilon_2 - \varepsilon_1)},$$

$$C_2 = C_1 \lambda_1'', \quad C_4 = C_3 \lambda_2'', \quad C_6 = C_5 \lambda_3''.$$

Finally, we find the bending moments, deflections and boundary forces in the arbitrary section of branches between nodes $i{-}1$ and i:

$$M_2(\bar{s}) = \frac{D_2}{d_1} \left(\alpha_2 \omega_{i'} + \beta_2 \omega_{i'-1} - \gamma_2 \frac{\psi_i}{d_1} + \eta_2 \frac{\psi_{i-1}}{d_1} \right) \cdot A_2(\bar{s}) - \frac{D_2}{u_2 d_1^2} \left(\gamma_2 \omega_{i'} + \right.$$

$$\left. \eta_2 \omega_{i'-1} - \mu_2 \frac{\psi_i}{d_1} + \xi_2 \frac{\psi_{i-1}}{d_1} \right) \cdot B_2(\bar{s}) + \frac{q + k_2 \cdot \psi_i}{u_2^2} \cdot C_2(\bar{s}) + \frac{k_2 \cdot \omega_{i'}}{u_2^2} \cdot D_2(\bar{s}),$$

$$y_2(\bar{s}) = \frac{u_2^2}{k_2} \left[-\frac{4 D_2}{d_1} \left(\alpha_2 \omega_{i'} + \beta_2 \omega_{i'-1} - \gamma_2 \frac{\psi_i}{d_1} + \eta_2 \frac{\psi_{i-1}}{d_1} \right) \cdot C_2(\bar{s}) + \right.$$

$$\frac{4 D_2}{u_2 d_1^2} \left(\gamma_2 \omega_{i'} + \eta_2 \omega_{i'-1} - u_2 \frac{\psi_i}{d_1} + \xi_2 \frac{\psi_{i-1}}{d_1} \right) \cdot D_2(\bar{s}) + \frac{q + k_2 \cdot \psi_i}{u_2^2} \cdot A_2(\bar{s}) +$$

$$\dots (3.151)$$

$$\left. \frac{k_2 \cdot \omega_{i'}}{u_2^3} \cdot B_2(\bar{s}) \right] - \frac{q}{k_2}, \quad M_1(\bar{s}) = \frac{D_1}{d_1} \left(\alpha_1 \omega_i + \beta_1 \omega_{i-1} - \gamma_i \frac{\psi_i}{d_1} + \eta_1 \frac{\psi_{i-1}}{d_1} \right) \times$$

$$A_1(\bar{s}) - \frac{D_1}{u_1 d_1^2}\left(\gamma_1\omega_i + \eta_1\omega_{i-1} - \mu_1\frac{\psi_i}{d_1} + \zeta_1\frac{\psi_{i-1}}{d_1}\right)\cdot B_1(\bar{s}) + \frac{k_1\psi_i}{u_1^2}\cdot C_1(s) +$$

$$\frac{k_1\cdot\omega_{i'}}{u_1^3}\cdot D_1(\bar{s}),$$

$$y_1(\bar{s}) = \frac{u_1^2}{k_1}\left[-\frac{4D_1}{d_1}\left(\alpha_1\omega_i + \beta_1\omega_{i-1} - \gamma_1\frac{\psi_i}{d_1} + \eta_1\frac{\psi_{i-1}}{d_1}\right)\cdot C_1(\bar{s}) +\right.$$

$$\left.\frac{4D_1}{u_1 d_1^2}\left(\gamma_1\omega_i + \eta_1\omega_{i-1} - \mu_1\frac{\psi_i}{d_1} + \zeta_1\frac{\psi_{i-1}}{d_1}\right)\cdot D_1(\bar{s}) + \frac{k_1\cdot\psi_i}{u_1^2}\cdot A_1(\bar{s}) + \frac{k_1\cdot\omega_{i'}}{u_1^3}\cdot B_1(\bar{s})\right],$$

$$\tag{3.152}$$

$$T_2(\bar{s}) = \frac{E\delta_2}{R_2}\cdot y_2(\bar{s}), \quad T_1(\bar{s}) = \frac{E\delta_1}{R_2}\cdot y_1(\bar{s}),$$

$$\bar{s} = s/d_1.$$

References

*Aleksandrov, A.V. 1963. Displacement method for analysing plate-beam structures. In: *Proc. Inst. Railway Transport Engineers*. Transzheldorizdat, Moscow. No. 174, pp. 20–28.

*Aleksandrov, A.V. 1965. Analysing cellular beam span structures by the displacement method. In: *Study of Theory of Structures*. Stroiizdat, Moscow. No. 14, pp. 209–213.

*Aleksandrov, A.V. 1968. Application of displacement method for plotting the force influence surfaces in plates of bridge plate-beam systems. In: *Study of Theory of Structures*. Stroiizdat, Moscow. No. 16, pp. 140–148.

*Aleksandrov, A.Ya. 1957. *Elastic Parameters of Ribbed Fillings of Three-layered Panels*. Novosibirsk, 144 pp.

*Aleksandrov, A.Ya., L.E. Bryukker, L.M. Kurshin and A.P. Prussakov. 1960. *Analysing Three-layered Panels*. Oborongiz, Moscow, 271 pp.

*Al'tenbakh, I. and V. Kissing. 1983. Constructing unidimensional finite elements based on Vlasov's semi-membrane theory of shells. In: *Dynamics and Strength of Machines*. Khar'kov. No. 38, pp. 32–36.

*Al'tenbakh, I. and V. Kissing. 1985. Constructing finite elements based on Vlasov's theory of shells. In: *Dynamics and Strength of Machines*. Khar'kov. No. 42, pp. 24–28.

*Antipov, A.A. 1964. Study of Bending and Stability of Three-layered Plates of Non-symmetrical Thickness with Allowance for Temperature Influences. Cand. Diss. Kiev.

*Biderman, V.L. 1965. Specific features of analysing thin-walled profiles for strength and stiffness. In: S.L. Ponomarev et al. (eds.) *Strength Analysis in Engineering*. Mashgiz, Moscow, Vol. 1, pp. 389–474.

*Biderman, V.L. 1977. *Mechanics of Thin-walled Structures. Statics*. Mashinostroenie, Moscow. 488 pp.

*Birman, S.E. 1961. Constrained torsion of thin-walled shafts with a closed rectangular profile and lateral diaphragms. *Izv. AN SSSR. OTN Mekhanika i Mashinostroenie*, no. 1, pp. 19–29.

*Bleikh, F. and E. Melan. 1938. *Equations in Finite Differences of Statics of Structures*. Department of Scientific and Technical Information, Khar'kov. 384 pp.

*Blokhina, I.V. and O.M. Ignat'eva. 1987. Analysing thin-walled cellular systems by the superelement method with spline-interpolation of displacements. Volgograd Engineering-Construction Institute, Volgograd. Deposited in VINITI, no. 1077, June 5, 1987.

*Bolotin, V.V., I.I. Gol'denblat and A.F. Smirnov. 1972. *Structural Mechanics. Present Status and Developmental Prospects.* Stroiizdat, Moscow.

*Borisov, V.I. 1969. Strength and Deformability of Three-layered Corrugations and Shells. Cand. Diss., Moscow.

*Borshchenko, N.K. 1971. Some Problems in the Bending of Non-symmetrical Three-layered Strips. Cand. Diss., Dnepropetrovsk.

*Bryukker, L.E. 1952a. Approximate solution of some problems of linear-lateral bending of three-layered strips with rigid orthotropic filler. In: *Analysis of Three-layered Panels and Shells.* Oborongiz, Moscow.

*Bryukker, L.E. 1952b. Linear-lateral bending of strips with rigid filler. In: *Analysis of Three-layered Panels and Shells.* Oborongiz, Moscow.

*Bryukker, L.E. 1969. Applicability range of the approximate theory of non-symmetrical three-layered strips. In: *Seventh All-Union Conference on the Thoery of Shells and Strips.* Nauka, Moscow.

*Bukalov, V.M. and A.A. Narusbaev. 1984. Design of Nuclear Submarines. Sudostroenie, Leningrad, 286 pp.

*Chudnovskii, V.G. 1952. Methods of Analysis of oscillations and stability of shaft systems, Kiev, 1952, 416pp.

*Efimov, A.I. 1962. Stability of Three-layered Cylindrical Shells with Crimp Filler under Combined Loading. Cand. Diss., Moscow.

*Ershov, V.V. 1961. Cylindrical bending of three-layered non-symmetrical strips with a lightweight filler. *Izvestiya Vuzov Aviatsionnaya Tekhnika*, No. 3.

*Gibshman, E.E. 1969. *Design of Metal Bridges.* Transport, Moscow. 248 pp.

*Gol'denveizer, A.L. 1949. Theory of thin-walled shafts. *Prikladnaya Matematika i Mekhanika*, vol. 13, no. 6.

*Grigolyuk, E.I. and P.P. Chulkov. 1973. *Stability and Oscillations of Three-layered Shells.* Mashinostroenie, Moscow.

*Ignatiev, V.A. 1973. *Analysing Periodic Shaft Systems.* Saratov State University. Saratov, 433 pp.

*Ignatiev, V.A. 1979. *Analysing Periodic Statically Indeterminate Shaft Systems.* Saratov State University. Saratov, 295 pp.

*Ignatiev, V.A. and T.M. Kaurova. 1989. Static analysis of cellular superelement systems using spline interpolation in forward and backward sweeps. Volgograd Engineering-Structural Institute, Volgograd. Deposited in VINITI, 11 July 1989, No. 4537.

*Il'yasevich, S.A. 1970. *Cellular Metal Bridges.* Transport, Moscow. 280 pp.

*Ivannikov, V.V. 1970. Optimum parameters of three-layered panels with ribbed and solid filers. *Trudy Gor'kovskogo Politekhn. In-ta*, No. 11, pp. 69–76.

*Karavanov, V.F. 1961. Bending and Stability of Three-layered Shells with Lightweight Fillers. Cand. Diss., Moscow Aviation Institute, Moscow.

*Kaurova, T.M. 1988. Static analysis of prismatic shells with multiconnected section and periodic structure. In: *Scientific-Field Conference on Spatial Structures.* Rostov-on-Don, pp. 57–58.

*Kirichenko, V.L. 1970. Stability and Natural Oscillations of Three-layered Strips Fixed Rigidly with Ribs. Cand. Diss., Dnepropetrovsk.

*Kiselev, V.A. 1935. *Beams and Frames on an Elastic Foundation.* Transzheldorizdat, Moscow, 43 pp.

*Kornoukhov, N.V. 1949. *Strength and Stability of Shaft Systems.* Stroiizdat, Moscow, 376 pp.

*Krylov, A.N. 1931. *Analysing Beams Resting on Elastic Foundation.* AN SSSR, Moscow.

154

*Kurshin, L.M. 1957. *Some Aspects of Bending and Stability of Three-layered Cylindrical Shells.* Cand. Diss., Institute of Mechanics, AN SSSR, Moscow.

*Kurshin, L.M. 1962. Review of work on analysing three-layered plates and shells. In: *Analysis of Spatial structures.* Gosstroiizdat, Moscow, No. 6, pp. 163–192.

*Kurshin, L.M. 1964. *Some Aspects of the Stability of Three-layered and Single-layered Plates and Shells at Normal and High Temperatures.* Doct. Diss., Novosibirsk.

*Kuznetsov, O.R. and V.V. Petrov. 1983. Application of the method of enlarged element for analysing straight closed prismatic shells produced from a ductile non-linear-elastic material. *Izvestiya Vuzov Stroitel'stvo i Arkhitektura,* no. 7, pp. 33–36.

*Lozhkin, O.B. 1976. *Some Aspects of Axisymmetrical Bending of Three-layered Rotation Shells.* Cand. Diss., Moscow.

*Luzhin, O.V. 1959. *Theory of Thin-walled Shafts with Closed Profile and Its Application to Bridge Construction.* Military Engineering Academy, Moscow.

*Maslennikov, A.M. and V.I. Pletnev. 1981. Designing the finite element method for multicellular systems with attenuations. *Stroitel'naya Mekhanika i Raschet Sooruzhenii,* no. 4, pp. 17–20.

*Meshcheryakov, V.B. 1964. Two articles on taking into account the impact of shear on a median surface. In: *Proc. Inst. Railway Transport Engineers,* no. 193.

*Meshcheryakov, V.B. 1965. Shear impact on the working of thin-walled shafts. *Inzh. Zh.,* vol. 5, no. 1.

*Mileikovskii, I.E. 1950a. Spatial covering of the stiff cellular shell type. In: *Studies on the Theory and Design of Thin-walled Structures* (V.Z. Vlasov, ed.). Stroiizdat, Moscow, pp. 39–42.

*Mileikovskii, I.E. 1950b. Some practical problems in analysing coverings of the cylindrical shell type. In: *Studies on the Theory and Design of Thin-walled Structures.* (V.Z. Vlasov, ed.). Stroiizdat, Moscow, pp. 232–258.

*Mileikovskii, I.E. 1960. *Design of Shells and Corrugations by the Displacement Method.* Gosstroiizdat, Moscow, 298 pp.

*Nemchinov, Yu.I. *Analysing Spatial Structures (Finite Element Method).* Budivel'nik, Kiev, 232 pp.

*Nemchinov, Yu.K. and A.V. Frolov. 1981. Design of buildings and structures by the spatial finite element method. *Stroitel'naya Mekhanika i Raschet Sooruzhenii,* no. 5, pp. 29–33.

*Nemchinov, Yu.K. and V.G. Kozyr'. 1984. Analysing thin-walled structures with allowance for the instantaneous stressed state by the spatial finite element method. *Stroitel'naya Mekhanika i Raschet Sooruzhenii,* no. 2, pp. 18–21.

*Obraztsov, I.F. 1956. *Design of Caisson-type Shells of Swept Wing Based on V.Z. Vlasov's Theory.* Proc. Moscow Aviation Institute, no. 59. Oborongiz, Moscow.

*Obraztsov, I.F. 1956. *Bending and Torsion of Multiple-sealed Caisson Structures.* Proc. Moscow Aviation Institute, no. 86. Oborongiz, Moscow, 68 pp.

*Pletnev, V.I. 1983. Analysing cellular systems by the method of forces and method of displacements together with the method of finite elements. In: *Structural Mechanics.* Leningrad Engineering Structural Institute, Leningrad, pp. 26–32.

*Postnov, V.A. (ed.) 1979. Substructuring Method for Analysing Structures. Sudostroenie, Leningrad, 288 pp.

*Postnov, V.A. and I.Ya. Kharkhurim. 1974. *Finite Element Method for Analysing Ship Structures.* Sudostroenie, Leningrad, 332 pp.

*Potapkin, A.A. 1984. *Design of Steel Bridges with Allowance for Plastic Deformations.* Transport, Moscow, 200 pp.

*Prussakov, A.P. 1957. *Stability and Natural Oscillations of Three-layered Anisotropic Strips with Filler.* Doct. Diss. Institute of Structural Mechanics, AN UkrSSR, Kiev.

*Prussakov, A.P. and N.K. Borshchenko. 1967. Bending of three-layered non-symmetrical strips with rigid filler. In: *Aerohydromechanics and Elasticity Theory.* Republic Interinstitutional Scientific Collection, no. 6, pp. 100–106.

*Rabinovich, I.M. 1921. *Application of the Theory of Finite Differences to Studying Continuous Beams*. Technical Committee of National Conference of Construction Industry, 96 pp.

*Segal, A.I. 1949. *High-rise Structures*. Stroiizdat, Moscow, 140 pp.

*Ship Building Mechanics—Reference. 1982. Sudostroenie, Leningrad. Vol 2, 464 pp.

*Smirnov, A.F., A.V. Aleksandrov, N.N. Shaposhnikov and B.Ya. Lashchennikov. 1964. *Analysing Structures Using Computers*. Stroiizdat, Moscow.

*Sokolov, O.L. 1967a. Bending of the walls of a composite tank under the impact of lateral hydrostatic pressure and axial loading. In: *30th Scientific Conference of Saratov Polytechnical Institute*, Saratov, pp. 84–92.

*Sokolov, O.L. 1967b. Some Aspects of Bending and Stability of Shaft Frames in an Elastic Medium, Composite Plates and Shells. Cand. Diss., Saratov, 200 pp.

*Sokolov, O.L. 1969. Bending of composite shells under impact of external pressure. In: *Theory of Analysis and Reliability of Instruments*. Proc. II Saratov Regional Conference of Young Scientists. Saratov State University, Saratov, pp. 52–58.

*Sokolov, O.L. 1970a. Cylindrical bending of three-layered strips with ribbed filler. In: *Construction and Theory of Structures*. Saratov, pp. 71–75.

*Sokolov, O.L. 1970b. Stability of a ring frame. In: *Design of Composite Systems of Coverings*. Scientific Proc. of Saratov Polytechnical Institute, Saratov, no. 46, pp. 43–55.

*Sokolov, O.L. 1984a. Design of multiconnected prismatic shells of periodic structure. *Stroitel'naya Mekhanika i Raschet Sooruzhenii*, no. 4, pp. 14–16.

*Sokolov, O.L. 1984b. Design of multiconnected prismatic shells of periodic structure in an elastic stage. In: *Studies on the Structural Mechanics of Shaft Systems*. Saratov Polytechnical Institute, Saratov, pp. 119–135. Deposited in VINITI, 12 Oct. 1984, No. 6661–84.

*Sokolov, O.L. 1984c. Statics of prismatic shells of a multiconnected section with a periodic structure. Vologodskii Polytechnical Institute, Vologda, 59 pp. Deposited in VINITI, 15 Jan. 1985, no. 402–85.

*Sokolov, O.L. 1986. Statics of prismatic shells of a multiconnected section with a periodic structure. Part 2. Analysing for the impact of mobile concentrated load on cellular span structures of bridges with a multiconnected section without diaphragms. Vologodskii Polytechnical Institute, Vologda, 94 pp. Deposited in VINITI, 14 Jan. 1986, No. 317-V.

*Sokolov, O.L. 1987a. Statics of prismatic shells of a multiconnected section with a periodic structure. Part 3. Analysing continuous cellular span structures of bridges with a multiconnected section without diaphragms for the impact of mobile concentrated load. Vologodskii Polytechnical Institute, Vologda, 43 pp. Deposited in VINITI, 5 March 1987, no. 1602-V 87.

*Sokolov, O.L. 1987b. Statics of cylindrical shells with a double wall and three-layered panels with ribbed filler of periodic structure. Vologodskii Polytechnical Institute, Vologda, 119 pp. Deposited in VINITI, 22 April 1987, No. 2806-V 87.

*Sokolov, O.L. 1988a. Statics of prismatic shells of a multiconnected section with a periodic structure. Part 4. Prismatic shells on an elastic foundation. Shells with double casings. Enlarged spatial element for digital analysis of multiconnected shells of medium length. Oscillations. Vologodskii Polytechnical Institute, Vologda, 87 pp. Deposited in VINITI, 22 April 1988, No. 3116–88.

*Sokolov, O.L. 1988b. Allowance for discrete location of ribs in problems of bending of three-layered strips. In: *Deformation and Failure of Structural Members and Material*. Mezhvuzovskii Sbornik, Leningrad-Vologda, pp. 16–21.

*Sokolov, O.L. 1991. Analysing thin-walled cylinders with a double wall as a shell of multiconnected section. In: *Long-time Strength of Building Materials and Aspects of Analysing Structural Members*. Mezhvuzovskii Sbornik, Leningrad-Vologda, pp. 48–51.

*Sokolov, O.L. 1992. Theory of analysing shells of multiconnected section with a periodic structure. In: *Republic Scientific Seminar on Mechanics and Technology of Polymer and Composite Materials and Structures, Abstracts*. St. Petersburg, 92 pp.

*Sokolov, O.L. and T.M. Kaurova. 1991. Spatial enlarged member for analysing prismatic shells of multiconnected section. In: *Long-time Strength of Building Materials and Aspects of Analysing Structural Members*. Mezhvuzovskii Sbornik, Leningrad-Vologda, pp. 39–44.

*Streletskii, N.S. et al. 1961. *Metallic Structures*. Gosstroiizdat, Moscow, 776 pp.

Strength, Stability, Oscillations—A Guide. Mashinostroenie, Moscow. Vol. 2, 464 pp.

Structural Mechanics in the USSR (1917–1967). 1969. Stroiizdat, Moscow, 423 pp.

*Timoshenko, S.P. 1955. *Stability of Elastic Systems*. Gostekhizdat, Moscow, 474 pp.

*Umanskii, A.A. 1939a. *Floating Bridges*. Transzheldorizdat, Moscow, 392 pp.

*Umanskii, A.A. 1939b. *Torsion and Bending of Thin-walled Aviation Structures*. Oborongiz, Moscow, 112 pp.

*Varvak, P.M., I.M. Buzun, A.S. Gorodetskii, V.G. Piskunov and Yu. N. Toloknov. 1981. *Finite Element Method*. Vyssh. Shkola, Kiev, 176 pp.

*Vasil'kov, B.S. 1950. Study of the spatial work of coatings and coverings of the two-layered prismatic shell type with a multiconnected section. In: *Studies on Aspects of Theory and Design of Thin-walled Structures* (V.Z. Vlasov, ed.). Moscow, pp. 197–212.

*Vasil'kov, B.S. 1970. Application of the method of finite elements in analysing shells, corrugations, ribbed and massive systems. In: *Theory and Design of Structures*. Proc. Central Institute of Building Structures. Moscow, no. 13, pp. 90–100.

*Vasitsina, T.N. 1962. Finite Deflections, Stability and Oscillations of Three-layered Non-symmetrical Shells. Cand. Diss., Moscow.

*Vlasov, V.Z. 1949. *General Theory of Shells and Its Application in Technology*. GITTL, Moscow, Leningrad, 784 pp.

*Vlasov, V.Z. 1958. *Thin-walled Spatial Systems*. Gosstroiizdat, Moscow, 502 pp.

*Vlasov, V.Z. 1959. *Thin-walled Elastic Shafts*. GIFML, Moscow, 568 pp.

*Vlasov, V.Z. and A.K. Morshchinskii. 1950. Contact problems in the theory of cylindrical shells fixed with linear ribs. In: *Studies on the Theory and Design of Thin-walled Structures* (V.Z. Vlasov, ed.). Moscow, pp. 170–175.

*Volk, S.I. 1967. Differential equilibrium equations of two elastic coaxial rotation shells joined with meridional stiffeners. In: *All-Union Conference on Statics and Dynamics of Thin-walled Spatial Structures, Abstracts*. Kiev.

*Volk, S.I. 1968. Method of Analysing Rotation Shells with Meridional Ribs Allowing for Cyclic Symmetrical Nature of Deformations. Cand. Diss., Kiev.

*Vol'nov, V.S. 1978. *Torsion of Cellular Bridge Spans*. Transport, Moscow, 136 pp.

*Vol'vich, S.I. 1939. Theory of the Stability of Shaft Systems. Doct. Diss., Moscow, 135 pp.

*Vol'vich, S.I. and O.L. Sokolov. 1968. Stability of axisymmetrical deformation of composite cylindrical shells with round ribs under axial compression. *Izvestiya Vuzov. Stroitel'stvo i Arkhitektura*, no. 1, pp. 27–32.

*Vyrbanov, Kh.P. 1980. General stability of thin-walled shafts with multiconnected closed contour of the cross-section. In: *Studies on the Theory of Structures*. Stroiizdat, Moscow, No. 24.

---

*Asterisked references are in Russian—General Editor

4

Condensation Methods in Problems of Free Oscillations and Stability

4.1 Introduction

A study of free oscillations and stability of complex multiple-membered structures using digital methods of structural mechanics involves solving algebraic eigenvector and eigenvalue problems of a high order of magnitude ($N \cong 10^3$ to 10^5). It follows that the accurate solution of such problems by standard methods is hardly feasible, if not impossible, even with sufficient computer resources.

The practical need to determine only a limited number of eigenvalues and corresponding vectors led to attempts to significantly reduce the magnitude of the problem and the development of methods for solving reduced sets involving eigenvalues and vectors. Methods of static and dynamic condensation based on reducing the order of the characteristic matrix by exchanging all secondary (auxiliary) degrees of freedom have gained considerable application in recent years (Guyan, 1965; Meisner, 1968; Araldsen, 1972; Wilson, 1974; Postnov, 1975; Dashevskii and Rotter, 1984*; Zemljanuchin (Zemlyanukhin), 1960; Zeinkiewicz 1967, 1975, 1975; Craig and Chaing-Jone Chang, 1976; Ignatiev and Chuban', 1984; Ki-Ook Kim and Anderson, 1984; Nemchinov and Kozyr', 1985; Leung, 1989; Tokuda and Sakata, 1989; Zhang, 1989; Zhuravleva and Gadyaka, 1989; Rosman, 1990; Dieker, 1991; Levy, 1991; Ignatiev, 1992; Belyi, 1993). Kammer et al. 1991, Bouhaddi et al. 1992, 1996, Suarez et al., 1992, Singh et al., 1992, Craig at al., 1968, Kim, 1985, 1995, Leung, 1978.

4.2 Statement of the Problem

The problem of finding eigenvalues and eigenvectors arising when studying free oscillations or the stability of a system described by a discrete analytical scheme is stated by a matrix equation of a general type

$$[C - \lambda I]\{z\} = 0. \qquad \qquad \dots (4.1)$$

---

*See Chapter 1 for these References—General Editor.

In a scalar form, the above equation represents a system of homogeneous linear equations (I is the identity matrix). A non-trivial solution of this system (i.e., $\{z\} \neq 0$) is possible only when the determinant of the matrix of equation coefficients (4.1) is equal to zero:

$$\det[C - \lambda I] = 0. \qquad \qquad \dots (4.2)$$

The matrix of coefficients $[C - \lambda I]$ of eqn. (4.1) is called the characteristic matrix of matrix C in eqn. (4.2) relative to λ, its characteristic equation. The roots of this equation are the eigenvalues $\lambda_i (i=1,2,\dots,N)$, all of which together are called the matrix spectrum.

When solving several problems of structural mechanics, including problems of free oscillations (ignoring damping and stability) using classical methods of structural mechanics, the stiffness or flexibility methods, the associated stiffness or flexibility matrices are symmetrical and except in unusual cases positive and definite. All of their elements are, therefore, real numbers. A consequence of this is that all the eigenvalues of eqn. (4.1) will be real numbers while eigenvectors form an orthonormal set.

When evaluating free oscillations utilising the displacement method (ignoring damping), matrix C in eqn. (4.1) is determined as:

$$C = M^{-1} K. \qquad \qquad \dots (4.3)$$

In problems of stability:

$$C = Y^{-1} K. \qquad \qquad \dots (4.4)$$

Here, K, M and Y are the stiffness, mass and loading potential matrices respectively.

When analysing in the form of the method of forces, matrix C is determined by the equation:

$$C = M\delta, \qquad \qquad \dots (4.5)$$

in which M is the diagonal matrix of mass and $\delta = K^{-1}$ is the flexibility influence coefficients matrix.

4.3 Static Condensation

The method of static condensation represents one of the most convenient and simple methods of reducing the unknowns in the substructuring method of solving dynamic problems (Guyan, 1965; Benfield, 1971; Wilson, 1974; Sapozhnikov, 1980; Shaposhnikov and Yudin, 1982; Gallager, 1984; and others*).

Let us study the process of deriving the stiffness and mass matrices of a superelement (substructure) by joining the members of the preceding level.

---

*See Chapter 1 for these References—General Editor.

The equation for free oscillations of the substructure (ignoring damping) will have the following form:

$$[K - \lambda M]\{z\} = 0, \qquad \qquad \dots (4.6)$$

in which $\lambda = \omega^2$ (ω is the cyclic frequency of oscillations).

All the degrees of freedom of the system under consideration (superelement) can be classified (dof) as primary (selected) or secondary (excluded).

Subscripts r and s are used to distinguish the displacement of nodes and the matrix elements of K and M corresponding to these degrees of freedom(DOF):

$$\left(\begin{bmatrix} K_{rr} & K_{rs} \\ K_{sr} & K_{ss} \end{bmatrix} - \lambda \begin{bmatrix} M_{rr} & M_{rs} \\ M_{sr} & M_{ss} \end{bmatrix} \right) \begin{Bmatrix} z_r \\ z_s \end{Bmatrix} = 0. \qquad \dots (4.7)$$

Assuming that the forces of inertia in the direction of the secondary degrees of freedom are negligible compared to the forces of the inertia of mass in the primary (selected) nodes, the second equation of system (4.7) can be simplified:

$$K_{sr} z_r + K_{ss} z_s = 0. \qquad \qquad \dots (4.8)$$

We find from this equation that

$$z_s = - K_{ss}^{-1} K_{sr} z_r. \qquad \qquad \dots (4.9)$$

This dependence establishes the relation between the primary and secondary unknowns in $\{z\}$.

Thus the displacement vector $\{z\}$ is represented in the form

$$\{z\} = \begin{Bmatrix} z_r \\ z_s \end{Bmatrix} = [A_r]\{z_r\}, \qquad \qquad \dots (4.10)$$

in which

$$[A_r] = \left[\frac{I_r}{-K_{ss}^{-1} K_{sr}} \right] - \qquad \qquad \dots (4.11)$$

is the matrix of static conversion (I_r is the unit matrix of the order of r).

Taking into consideration eqn. (4.10), the following equations for statistically condensed stiffness and mass matrices of the substructure (superelement) can be derived (Argiris, 1961):

$$\left[K_{rr}^{\bullet} \right] = [A_r]^T [K][A_r] = K_{rr} - K_{rs} K_{ss}^{-1} K_{sr}. \qquad \dots (4.12)$$

$$\left[M_{rr}^{\bullet} \right] = [A_r]^T [M][A_r] = M_{rr} - M_{rs} M_{ss}^{-1} M_{sr}.$$

Solving the problem further

$$\left(\left[K_{rr}^* \right] - \lambda \left[M_{rr}^* \right] \right) \{z\} = 0, \qquad \ldots (4.13)$$

from which the frequencies and forms of free oscillations of the superelement can be found. By carrying out the conversion (4.10) a few times until the final level of the superelement is reached, i.e., to the level of the structure as a whole, solving eqn. (4.13) corresponding to this level will give the first 'r' eigenvalues (frequencies) and eigenvectors (modes of oscillations).

The major drawback of this method is the analytical error arising from the assumption that the inertial forces in the secondary nodes are negligible. Another drawback is that the accuracy of the result depends on the selection of the condensed nodes (main degrees of freedom).

Another modification of reduction is possible by ignoring in eqn. (4.7) the static forces, i.e., the stiffness associated with the secondary degrees of freedom, compared to the forces of inertia (Egeland, 1974). In this case, instead of eqn. (4.8), we obtain:

$$M_{sr} z_r + M_{ss} z_s = 0, \qquad \ldots (4.14)$$

that is

$$z_s = - M_{ss}^{-1} M_{sr} z_r. \qquad \ldots (4.15)$$

Then,

$$\{z\} = \begin{Bmatrix} z_r \\ z_s \end{Bmatrix} = \left[A_r^{(M)} \right] \{z_r\}, \qquad \ldots (4.16)$$

in which

$$\left[A_r^{(M)} \right] = \left[\frac{I_r}{-M_{ss}^{-1} M_{sr}} \right]. \qquad \ldots (4.17)$$

Using matrix $A_r^{(M)}$, we find that

$$\left[M_{rr}^* \right] = \left[A_r^{(M)} \right]^T [M] \left[A_r^{(M)} \right] = M_{rr} - M_{rs} M_{ss}^{-1} M_{sr},$$

$$\left[K_{rr}^* \right] = \left[A_r^{(M)} \right]^T [K] \left[A_r^{(M)} \right] = K_{rr} - K_{rs} M_{ss}^{-1} M_{sr} - \qquad \ldots (4.18)$$

$$M_{rs} M_{ss}^{-1} M_{ss} + M_{rs} M_{ss}^{-1} K_{ss} M_{ss}^{-1} M_{sr}.$$

This approach too suffers from the same drawback as the preceding one: the possibility of large errors in this case due to ignoring the static forms of deformations in the equilibrium equations associated with the secondary nodes.

By combining these two approaches, as suggested by Benfied, W.A. and R.F. Hruda, 1972; see Chapter 1, the following relation is obtained:

$$[A]=(1-\alpha)[A_r]+\alpha[A_r^{(M)}] \quad \text{at} \quad (0 \le \alpha \le 1). \qquad \text{... (4.19)}$$

Analyses carried out by Benfield (see Chapter 1) with α varying from zero to 0.4 showed that the higher the value of α, the more accurate the high frequency values and the greater the error of calculating the low frequencies.

For low frequencies, Benfield therefore recommends $\alpha = 0$, which corresponds to the method of total static condensation.

4.4 Dynamic Condensation

Method of synthesising oscillation forms (modal synthesis) of substructures
This method has been discussed in several works (Ivanteev et al., 1984, Hurty, 1965, Araldsen, 1970, 1972 see Chapter 1). Its algorithm is as follows.

Displacements of the secondary (slave) nodes of the substructure are represented as the sum of their static displacements caused by displacements of the primary nodes and displacements of firmly fixed primary nodes in the substructure $(\{z_r\} = 0)$ represented as resolutions of natural forms of oscillations, i.e.,

$$z_s = z_s^{st} + z_s^{dyn} = z_s^{st} + \Phi_s q_s, \qquad \text{... (4.20)}$$

in which z_s^{st} is determined using eqn. (4.9); and q_s represents the generalised co-ordinates (amplitudes); $\Phi_s = \left(\Phi_s^{(1)}, \Phi_s^{(2)}, \ldots, \Phi_s^{(N)}\right)$, the natural forms of oscillations; N the number of sustained forms of oscillations.

Then, the vector of the displacements of the substructure can be represented as follows:

$$\{z\} = \begin{Bmatrix} z_r \\ z_s \end{Bmatrix} = [A_D] \begin{Bmatrix} z_r \\ q_s \end{Bmatrix} \qquad \text{... (4.21)}$$

in which

$$[A_D] = \begin{bmatrix} I_r & 0 \\ -K_{ss}^{-1} K_{sr} & \Phi_s \end{bmatrix} \qquad \text{... (4.22)}$$

is the matrix of dynamic conversion Φ_s the matrix of the natural forms of the substructure oscillations with the main nodes fixed $(\{z\}_r = 0)$ and derived by solving the following problem:

$$\left[K_{ss} - \lambda^{(s)} M_{ss}\right]\{z_s\} = 0. \qquad \text{... (4.23)}$$

The total accounting of the dynamic properties of inner nodes of the substructure depends on the number of eigenvectors Φ_s included in matrix

---

*See chapter 1 for these References—General Editor.

$[\Phi_s]$. If it contained all the eigenvectors, there is total accounting of the dynamic properties of the secondary nodes of the substructure. However, usually to condense the volume of computational work, only n_s initial forms of oscillations out of N_s are included in matrix $[\Phi_s]$, i.e., $n_s < N_s$.

Using the conversion matrix (4.22), we obtain the following equation for the stiffness and mass matrices of the substructure on transition to co-ordinates z_r and q_s:

$$[K_D] = [A_D]^T [K][K][A_D] \begin{bmatrix} K_{rr}^* & 0 \\ 0 & \Lambda_s \end{bmatrix}, \qquad \qquad \ldots (4.24)$$

$$[M_D] = [A_D]^T [M][A_D] \begin{bmatrix} M_{rr}^* & M_{rq} \\ M_{qr} & M_{Ns} \end{bmatrix},$$

in which M_{rr}^* and K_{rr}^* are the statically condensed stiffness and mass matrices determined using eqn. (4.12); A_s the diagonal matrix of the initial n_s of N_s eigenvalues obtained by solving problem (4.23); and

$$M_{qr} = M_{rs}\Phi_s - K_{rs}K_{ss}^{-1} M_{ss} \Phi_s; \qquad M_{qr} = M_{rq}^T.$$

In order to eliminate co-ordinate q_s, the second row of matrix equation for natural oscillations of the substructure is used:

$$\left[K_D - \omega^2 M_D\right] = \{z\}, \qquad \qquad \ldots (4.25)$$

which has the following form:

$$\Lambda_s q_s = \Lambda M_{qr} z_r - \Lambda I_{n_s} q_s = 0,$$

in which $\Lambda = \omega^2$ is the diagonal matrix whose order is taken as equal to n_s.

Hence,

$$q_s = B_s z_r$$

in which

$$B_s = (\Lambda_s - \Lambda)^{-1}$$

Thus,

$$\{z\} = [B_D]\{z_r\}, \text{ in which } [B_D] = \begin{bmatrix} I_r \\ M_s \end{bmatrix}. \qquad \qquad \ldots (4.26)$$

Using this conversion matrix, the equation is derived for the stiffness and mass matrices of the substructure condensed to the primary degrees of freedom:

$$[\tilde{K}_{rr}] = [B_D]^T [K_D][B_D], \qquad \qquad \ldots (4.27)$$

$$\left[\tilde{M}_{rr}\right] = \left[B_D\right]^T \left[\tilde{M}_D\right]\left[B_D\right].$$

Free oscillations Eqn. (4.25) for the substructure will now contain only the primary degrees of freedom:

$$\left[\tilde{K}_{rr} - \omega^2 \tilde{M}_{rr}\right]\{z_r\} = 0. \qquad \qquad \text{... (4.28)}$$

The free oscillations equation for the entire structure will be:

$$\sum_t \left[\tilde{K}_{rr}^{(t)} - \omega^3 \tilde{M}_{rr}\right]\{z_r^{(t)}\} = 0. \qquad \qquad \text{... (4.29)}$$

Here, summation is done for all the t substructures figuring in the structure under consideration.

From the practical point of view, the most interesting case is $1 < n_s < N_s$ when incomplete consideration of the dynamic properties of the structure is coupled with a significant reduction of the order of the characteristic eqn. (4.2).

To compare the results obtained by static and dynamic condensation methods, let us consider the example of analysis cited by Blokhina (1989; see Chapter 2).

Example 4.1. A square strip of isotropic material hinged along the contour has been divided into 16 FE by a nodal grid. The weights according to the algorithm of the finite element method are assigned to the nodes

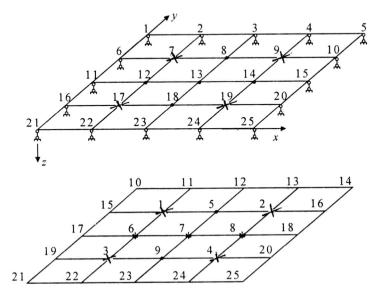

Fig. 4.1

Table 4.1

Sl. No. i	ω_i by finite elements method	Static condensation	Dynamic condensation, $n_s = 2$	Dynamic condensation, $n_s = 3$	Dynamic condensation, $n_s = 4$
1	4.803676	4.832184	4.83220	4.832194	4.803669
		(0.59)	(0.6)	(0.59)	(0.0004)
2	11.703461	12.612890	11.703504	11.703504	11.703804
		(7.8)	(0.0004)	(0.0004)	(0.0004)
3	11.703461	12.612890	11.703504	11.703504	11.703804
		(7.8)	(0.0004)	(0.0004)	(0.0004)
4	17.361279	—	17.361394	17.361394	17.361394
		—	(0.0007)	(0.0007)	(0.0007)
5	21.85413	21.854839	21.854839	21.854839	21.854244
		(20.56)	(0.0032)	(0.0033)	(0.0005)
6	21.85413	—	—	21.854862	21.854244
		—	—	(0.0033)	(0.0005)
7	25.32467	25.228238	25.228238	25.324731	25.324731
		(15.43)	(0.0036)	(0.0002)	(0.0002)
8	25.32467	—	25.324751	25.324731	25.324731
		—	(0.0002)	(0.0002)	(0.0002)
9	30.422153	—	—	—	30.422278
		—	—	—	(0.0003)

Note: The values shown in parentheses are the deviations in percentages from those of the finite element method adopted as accurate.

(point loads concentrated in the nodes). The numeration of nodes is shown in Fig. 4.1, a (dots depict the condensation nodes i.e. main and crosses the secondary i.e., slave nodes).

Using renumeration matrix $\left[\tilde{I}\right]$ whose structure is seen clearly from eqn. (1.9), the degrees of freedom in the stiffness and mass matrices of the structure under consideration are renumbered such that the linear and angular displacements are distinguished:

$$\left[\tilde{K}\right] = \left[\tilde{I}\right]\left[K\right]\left[\tilde{I}\right]^T = \begin{bmatrix} K_{zz} & K_{z\varphi} \\ K_{\varphi z} & K_{\varphi\varphi} \end{bmatrix},$$

$$\left[\tilde{M}\right] = \left[\tilde{I}\right]\left[M\right]\left[\tilde{I}\right]^T = \begin{bmatrix} M_{zz} & 0 \\ 0 & 0 \end{bmatrix},$$

in which M_{zz} is the diagonal mass matrix.

By static condensation using eqns. (4.11) and (4.12) and excluding the rotation angles, i.e., massless degrees of freedom, we arrive at the problem:

$$\left[M_{z,z}^{-1}\tilde{K}_{z,z} - \lambda I\right] = 0,$$

in which

$$\tilde{K}_{z,z} = K_{z,z} - K_{z,\varphi}K_{\varphi,\varphi}^{-1}K_{\varphi,z} -,$$

is the statically condensed stiffness matrix.

Solution of this problem gives eigenvalues $\lambda_i = \omega_i^2$ which are standard (accurate) for the adopted FE grid of the plate. The values of calculated frequencies ω_i are shown in the second column of Table 4.1.

Let us again do the static condensation after eliminating the secondary nodes shown in Figure 4.1,a in the form of crosses and renumbering the degrees of freedom (nodes) in the same order as shown in Figure 4.1,b. Later, using matrix $\left[\tilde{E}_1\right]$, derived as in eqn. (4.9), the statically condensed matrices are found:

$$\left[\tilde{K}_i\right] = \left[\tilde{I}_1\right]\left[\tilde{K}\right]\left[\tilde{I}_1\right]^T,$$

$$\left[\tilde{M}_i\right] = \left[\tilde{I}_1\right]\left[\tilde{M}\right]\left[\tilde{I}_1\right]^T, \qquad \qquad \dots (4.30)$$

To derive the dynamically condensed matrices using eqns. (4.24) to (4.26), problem (4.23) has to be first solved for the structure with fixed primary nodes and free secondary ones (depicted as crosses in Fig. 4.1). In this case, four mode shapes are available for oscillations with the main nodes fixed (condensation nodes) (Fig. 4.2).

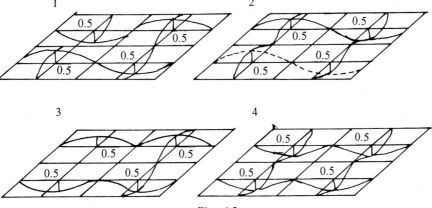

Fig. 4.2

By using the condensed matrices $\tilde{K}_1$ and $\tilde{M}_1$ and completing their construction according to eqns. (4.24) to (4.26), we obtain the dynamically condensed stiffness and mass matrices of the structure.

Table 4.1 shows the effect of the number of oscillation forms of the secondary nodes contained in the analysis on the accuracy of determining the natural frequencies. The frequency values are shown with common multiplier $l^{-2}\sqrt{(D/m)}$ (l is the length of the side of the square plates, D the plate stiffness and m the value of the point mass in the nodes of the FE grid).

Analysis shows that, at $n_s = N_s = 4$, there is total accounting of the dynamic properties of all N_s values of eliminable secondary nodes and the entire frequency spectrum is for practical purposes accurately determined. In this case, eqn. (4.21) represents the substitution of some variables by others while maintaining the total number of generalised displacements.

At $n_s = 0$, eqn. (4.21) is transformed into (4.10) which corresponds to a case of static condensation.

Analytical results show that at $n \leq 3$, the value of the first frequency $\omega_1 = \omega_{\min}$ barely differs from the value calculated by the static condensation method. Therefore, use of the dynamic condensation method for determining only the minimum oscillations frequency is not desirable due to notable consumption of machine time compared with the static condensation method.

Since the number of the main degrees of freedom is small compared with their total ($n \ll N$), solving the problems of eigenvalues and eigenvectors (4.23) for a substructure with matrix order $(N - n) \cdot (N - n)$ similarly endangers many difficulties. Therefore, multilevel dynamic condensation is recommended (Ivanteev, 1984 see Chapter 2); it represents a modification of the method of component-wise synthesis of forms (Bate and Wilson, 1976; 1982, pp. 341–344; see Chapter 2). Further, for each substructure of the lowest level (when analysing by the finite element displacement method) fixed in condensation nodes, the incomplete problem of eigenvalues and eigenvectors is solved by some standard method (e.g., by the method of simultaneous iterations in subspace). The process of constructing substructures of an increasingly higher level and solving their incomplete problems of eigenvalues and eigenvectors is continued until the substructures join to form the total structure.

The main drawbacks of this method are the complexity of the algorithm and the possibility of application only to the equation of the finite element displacement method.

Another variant of the dynamic condensation method, called the frequency-dynamic condensation method, has been discussed by Grinenko and Mokeev (1988). In this method, the auxiliary degrees of freedom are eliminated on the basis of the initial matrix equation, i.e., accurately, while the condensation of stiffness and mass matrices is done by equating the main frequencies of the condensed system with the basic frequencies of the original system.

4.5 Frequency-dynamic Condensation

As in the static condensation method, let us represent eqn. (4.6) in the form of eqn. (4.7).

By eliminating the auxiliary unknowns z_s, we obtain the matrix equation condensed to the basic unknowns:

$$\left[\left(K_{rr} - \lambda M_{rr}\right) + D_{rr}^{(s)}(\lambda)\right](z_r) = 0, \qquad \qquad \text{... (4.31)}$$

in which

$$D_{rr}^{(s)}(\lambda) = -\left(K_{ss} - \lambda M_{rs}\right)\left(K_{ss} - \lambda M_{ss}\right)^{-1}\left(K_{sr} - \lambda M_{sr}\right). \qquad \text{... (4.32)}$$

Let eqn. (4.31) be represented as

$$\left[\left(K_{rr} - \lambda M_{rr}\right) + \left(K_{rr}^{(s)} - \lambda M_{rr}^{(s)}\right)\right](z_r) = 0, \qquad \qquad \text{... (4.33)}$$

in which $K_{rr}^{(s)}$ and $M_r^{(ss)}$ can be found from the condition that the eigenvalues λ of the initial eqn. (4.6) and of the condensed equation in forms (4.31) and (4.33) coincide.

Then, using the two eigenvalues of eqn. (4.6) and equating eqns. (4.31) and (4.32), the following system of equations is obtained:

$$K_{rr}^{(s)} - \lambda_{1,rs}\, M_{rr}^{(s)} = D_{rr}^{(s)}\left(\lambda_{1,rs}\right),$$

$$K_{rr}^{(s)} - \lambda_{2,rs}\, M_{rs}^{(s)} = D_{rr}^{(s)}\left(\lambda_{2,rs}\right), \qquad \qquad \text{... (4.34)}$$

in which $\lambda_{1,rs}$ and $\lambda_{2,rs}$ are the limiting eigenvalues (λ_{max} and λ_{min}) of eqn. (4.6).

From this, we find:

$$M_{rr}^{(s)} = \frac{1}{\left(\lambda_{2,rs} - \lambda_{1,rs}\right)}\left[D_{rr}^{(s)}\left(\lambda_{1,rs}\right) - D_{rr}^{(s)}\left(\lambda_{2,rs}\right)\right],$$

$$K_{rr}^{(s)} = \frac{1}{\left(\lambda_{2,rs} - \lambda_{1,rs}\right)}\left[\lambda_{2,rs}\, D_{rr}^{(s)}\left(\lambda_{1,rs}\right) - \lambda_{1,rs}\, D_{rr}^{(s)}\left(\lambda_{2,rs}\right)\right]. \qquad \text{... (4.35)}$$

The limiting eigenvalues ($\lambda_{1,rs}$ and $\lambda_{2,rs}$) and the natural frequencies corresponding to them (basic frequencies or condensation range in the terminology of Grinenko and Mokeev) are fixed on the basis of the available approximate evaluations or analyses. The approximation accuracy of dynamic conversion matrix (4.33) depends on the assigned values of these basic frequencies, this being a serious drawback of this method.

An advantage of this method is the possibility of stepwise elimination of secondary degrees of freedom and evaluation of their correct assignment as secondary ones based on comparing the approximation errors of the dynamic conversion matrix in the condensation range.

In the frequency-dynamic condensation method, it has been assumed that the main frequencies are close to the limiting ones. However, analyses carried out by Blokhina (1987) showed that such an approach to selecting $\omega_{1.0}$ and $\omega_{2.0}$ gives in several cases approximate results due to unsatisfactory selection of the condensation nodes. The authors of this method, Grinenko ad Mokeev (1985, 1988) therefore suggest a special algorithm of stepwise elimination of secondary degrees of freedom and evaluating the correctness of their classification as secondary ones based on comparing the errors of approximation of the dynamic conversion matrix in the condensation range.

The labour as well as the effectiveness of this algorithm can be affected if the region between the main frequencies $\omega_{1.0}$ and $\omega_{2.0}$ is sufficiently narrowed, i.e., by setting up the problem of finding only a small part of the frequency spectrum or any frequency of interest, this being very important for solving practical problems. Such an approach is most effective when improving the accuracy levels of frequency values, e.g., by the static condensation method. In the modified frequency-dynamic condensation method suggested by us, matrix $D_{rr}^{(s)}(\lambda)$ is approximated by matrix $\left(K^{(s)} - \lambda M_{rr}^{(s)} \right)$ not in the entire condensation range as shown by Grinenko and Mokeev (1985), but in a small interval of each of the frequencies under analysis.

To evaluate the accuracy of results obtained by the modified frequency-dynamic condensation method, an example is discussed below.

Example 4.2. Let us examine again the plate shown in Fig. 4.1, a. Finite element analysis is carried out using the common mass matrix. Table 4.2 shows the values of the frequencies of the first five modes of oscillations of the plate for the initial node and those reduced to nodes 5, 6, 7, 8 and 9 (Fig. 4.1,b) of the nodal mesh obtained by the methods of static condensation, dynamic condensation and modified frequency-dynamic condensation.

The frequency values in Table 4.2 have been given without multiplier $l^{-2}\sqrt{(D/\overline{m}_0)}$, in which $\overline{m}_0$ is the mass per unit area.

Table 4.2

Frequency, ω_i	Accurate solution	Finite element method, 16×16	Error, Δ, %	Static condensation	Δ, % of finite element method
ω_1	19.739	19.29	2.27	19.357	0.3
ω_2	49.348	47.81	3.12	49.479	3.3
ω_3	49.348	47.81	3.12	49.479	3.3
ω_4	78.956	73.48	6.94	84.539	13.0
ω_5	98.696	96.459	2.27	97.945	1.5

Table 4.2 concld.

Frequency, ω_i	Dynamic condensation $n_s = 1$	Δ, % of finite element method	Dynamic condensation ($n_s = 2$)	Δ, % of finite element method	Modified frequency-dynamic condensation	Δ, % of finite element method
ω_1	19.47	0.93	19.40	0.57	19.285	0.026
ω_2	49.33	3.1	48.316	1.05	47.82	0.02
ω_3	49.33	3.1	48.316	1.05	47.82	0.02
ω_4	77.851	5.9	77.851	5.9	74.79	1.75
ω_5	97.195	0.76	96.498	0.04	96.468	0.01

4.6 Frequency-dynamic Condensation for Equations of General Type (4.1)

As demonstrated in section 4.2, when analysing in the form of the method of forces, an equation of the following type is resolved:

$$[M\delta - \lambda I]\{z\} = 0, \qquad \dots (4.36)$$

in which M is the diagonal matrix of mass and δ the flexibility matrix of the structure.

After separating the degrees of freedom of the structure into primary and secondary ones, eqn. (4.36) can be expressed as follows:

$$\left[\begin{bmatrix} \delta_{rr} & \delta_{rs} \\ \delta_{sr} & \delta_{ss} \end{bmatrix} \begin{bmatrix} M_r & 0 \\ 0 & M_s \end{bmatrix} - \lambda \begin{bmatrix} I_r & 0 \\ 0 & E_s \end{bmatrix} \right] \begin{Bmatrix} z_r \\ z_s \end{Bmatrix} = 0 \qquad \dots (4.37)$$

in which index 'r' refers to the primary and index 's' to the secondary degrees of freedom and I_r and I_s are the unit matrices of the corresponding orders.

The relation between the primary (z_r) and secondary (z_s) variables obtained from the second equation of system (4.37) is as follows:

$$z_s = -\left(\delta_{ss} M_s - \lambda I_s \right)^{-1} \delta_{sr} M_r z_r. \qquad \dots (4.38)$$

Substituting the above in the first equation, we obtain

$$\left[\left(\delta_{rr} M_r - \lambda I_r \right) + D_{rr}^{(s)}(\lambda) \right] (z_r) = 0, \qquad \qquad \text{... (4.39)}$$

in which

$$D_{rr}^{(s)}(\lambda) = -\delta_{rs} M_s \left(\delta_{ss} M_s - \lambda I_s \right)^{-1} \delta_{sr} M_r \qquad \text{... (4.40)}$$

is the dynamic conversion matrix of auxiliary displacements.

Matrix $D_{rr}^{(s)}$ can be approximated in the primary frequency range (in the condensation range) by

$$\tilde{D}_{rr}^{(s)}(\lambda) = \delta_{rr} M_r^{(s)}, \qquad \qquad \text{... (4.41)}$$

in which matrix $M_r^{(s)}$ will have the physical sense of the dynamic mass transport matrix M_s from auxiliary nodes 's' into primary 'r'. Coefficients of matrix $M_r^{(s)}$ are obtained from the conditions of the matrices $D_{rr}^{(s)}(\lambda)$ and $\tilde{D}_{rr}^{(s)}(\lambda)$ coinciding in one of the given condensation frequency ranges, e.g. minimal:

$$D_{rr}^{(s)}\left(\lambda_{\max(r,s)} \right) = \tilde{D}_{rr}^{(s)}\left(\lambda_{\max(r,s)} \right), \qquad \text{... (4.42)}$$

in which $\lambda_{\max(r,s)} = \omega_{\min}^{-2}$ is the upper value of the condensation range calculated approximately by any known method, e.g., by the static condensation method.

We find from eqn. (4.42) that

$$M_r^{(s)} = \delta_{rr}^{-1} \tilde{D}_{rr}^{(s)}\left(\lambda_{\max(r,s)} \right). \qquad \qquad \text{... (4.43)}$$

Equation (4.39) with allowance for eqn. (4.42) can be represented as

$$\left[\delta_{rr} M_r + D_{rr}^{(s)}\left(\lambda_{\max(r,s)} \right) - \lambda I_r \right] \{ z_r \} = 0, \qquad \text{... (4.44)}$$

or

$$\left[\delta_{rr} \left(M_r + M_r^{(s)} \right) - \lambda I_r \right] \{ z_r \} = 0.$$

By solving the characteristic equation

$$\left[\delta_{rr} \left(M_r + M_r^{(s)} \right) - \lambda I_r \right] = 0, \qquad \qquad \text{... (4.45)}$$

we find all of its eigenvalues and the eigenvectors corresponding to them.

4.7 Block Condensation Based on Equivalence of Partial Eigenvalues

The main drawback of the above frequency-dynamic condensation method for eqn. (4.36) is the need to find first the value of parameter λ_{max} by some method and convert the matrices of high orders in eqn. (4.38) at high values of N and low values of n (N is the total number of degrees of freedom and n the number of the main degrees of freedom).

Therefore, when the number of secondary degrees of freedom is large $(s = N - n = 10^2$ to $10^4)$, all of them are subdivided into t groups, i.e., the entire characteristic matrix $[C - \lambda I]\{z\} = 0$ is presented in a block form of order $(t + 1)\cdot(t + 1)$ blocks.

By isolating from these t groups the group s, the oscillations equation for the conditions of partial system can be represented in the form of eqn. (4.37). By solving this equation, we can find the corresponding partial eigenvalues. By selecting from these values λ_{mas}, dynamic condensation of mass in secondary nodes of group 's' to nodes of the primary group 'r' can be carried out using eqn. (4.43).

After dynamic conversion of all the t groups of masses, we obtain the final equation for the condensed system:

$$\left| \delta_{rr} \left(M_r + \sum_{s=1}^{t} M_r^{(s)} \right) - \lambda I_r \right| = 0, \qquad \ldots (4.46)$$

in which summation is carried out for all the t groups into which the auxiliary degrees of freedom have been divided.

Another sequence of block condensation is also possible, e.g., when the condensation in the i-th step is carried out to the condensed masses of the preceding $(i-1)$-th condensation stage.

Then, in the i-th condensation stage, eqn. (4.45) will assume the following form:

$$\left| \delta_{rr} \left(\tilde{M}_r^{(i-1)} + \tilde{M}_r^{(s,i)} \right) - \lambda I_r \right| = 0, \qquad \ldots (4.47)$$

in which $\tilde{M}_r^{(i-1)} + \tilde{M}_r^{(i-2)} + \tilde{M}_r^{(s,i-1)}$.

In a partial case when the condensation is carried out to a system with a single degree of freedom, the system of eqns. (4.37) should be regarded as a matrix system of two linear homogeneous algebraic equations. By solving this system, we find:

$$\lambda_{max} = \frac{1}{2}\left(M_r \delta_{rr} + M_s \delta_{ss} \right) +$$

$$\sqrt{\frac{1}{4}\left(M_r \delta_{rr} + M_s \delta_{ss} \right) - M_r M_s \left(\delta_{rr} \delta_{ss} - \delta_{rs} \delta_{sr} \right)}. \qquad \ldots (4.48)$$

By substituting this result in eqns. (4.43) and (4.40) and, after conversions, we obtain the equation for mass during its transport from node 's' into node 'r':

$$M_r^{(s)} = -\frac{\delta_{rs}\,\delta_{sr}\,M_s\,M_r}{\delta_{rr}\left(\delta_{ss}\,M_s - \lambda_{\max(r,s)}\right)}. \qquad \ldots (4.49)$$

By introducing the concept of the mass condensation coefficient to the point of mass application M_r, we find

$$K_{r,s} = \frac{M_r^{(s)}}{M_s} = -\frac{\delta_{rs}\,\delta_{sr}\,M_r}{\delta_{rr}\left(\delta_{ss}\,M_s - \lambda_{\max(r,s)}\right)}. \qquad \ldots (4.50)$$

Another variant of the equation for condensation coefficient can be obtained by representing eqn. (4.39), after substituting in it eqn. (4.41), in the form:

$$\delta_{rr}\,M_r - \lambda_{\max(r,s)} + \delta_{rr}\,K_{r,s} - M_s = 0. \qquad \ldots (4.51)$$

Thus, we find

$$K_{r,s} = \frac{1}{\delta_{rr}\,M_s}\left(\lambda_{\max(r,s)} + \delta_{rr}\,M_r\right). \qquad \ldots (4.52)$$

Then, eqn. (4.46) can be represented in the following form for the partial case under consideration:

$$\left|\delta_{rr}\,M_r\left(1 + \sum_{s=1}^{N} K_{r,s}\right) - \lambda\right| = 0, \qquad \ldots (4.53)$$

in which N stands for the total number of masses in the system.

We thus derive the approximate equation for $\lambda_{\max}$ of initial eqn. (4.1) during its condensation to an equation with a single unknown:

$$\lambda_{\max} = \delta_{rr}\,M_r\left(1 + \sum_{\substack{s=1 \\ (s\neq r)}}^{N} K_{r,s}\right). \qquad \ldots (4.54)$$

Here, index 'r' pertains to the number of the node (degree of freedom) to which all the masses of the system are condensed and M_r is the mass found in the given node. This equation wholly agrees with that given by Zemljanuchin (Zemlyanukhin) (1960) and Zemlyanukhin (1962, 1963).

Example 4.3. To illustrate the degree of accuracy of eqns. (4.50), (4.52) and (4.54), let us study natural oscillations of a beam of constant section hinged at the ends and carrying equidistantly placed point masses $M_1 = M_2 = M_3 = M_4 = M_5 = M$ (Fig. 4.3).

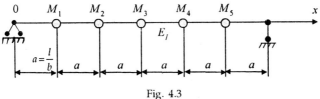

Fig. 4.3

The accurate values of parameter $\lambda = \omega^{-2}$ for this system have been determined by equations given by Ignatiev and Ignat'eva (1987, p. 130; Chapter 2):

$$\lambda_k = \frac{Ma^3}{48EJ} \cdot \frac{2 + \cos\dfrac{k\pi}{N+1}}{\sin^4\dfrac{k\pi}{2(N+1)}}, \qquad \ldots (4.55)$$

in which k is the number of the eigenvalue and N the number of masses (number of degrees of freedom).

For $k = 1$, $N = 5$ and $a = l/6$, we obtain $\lambda_{max} = 0.06181\ Ml^3/EI$. Considering that in a case

$$\delta_{11} = \delta_{55} = 25b,\ \delta_{22} = \delta_{44} = 64b,\ \delta_{33} = 81b,\ \delta_{31} = \delta_{13} = \delta_{35} = \delta_{53} = 39b,$$

$$\delta_{32} = \delta_{23} = \delta_{34} = \delta_{43} = 69b \quad \left(b = l^3/3888EJ\right)$$

according to eqns. (4.50 or 4.52) and (4.54), we find

$$\lambda_{max} = 0.06225\ Ml^3/EJ.$$

The above value differs from the accurate solution by 1.12%. The error of frequency is 0.563%. Use of eqns. (4.50) and (4.52) gives the same numerical results.

Example 4.4. Let us determine the fundamental frequency of natural oscillation of a bevelled rhombic isotropic plate studied in *Substructuring Method in Structural Analysis* (Maslennikov 1987, pp. 168–172; Chapter 2) (Fig. 4.4).

For this plate at $v = 0.3$, the following matrix $C = K^{-1}M$ was obtained with allowance for the symmetry of the system in Maslennilov; 1987, p. 172; Chapter 2):

$$C = \frac{\overline{m}_0 l^3}{D} \begin{bmatrix} 0.280285l & 0.127785l & -0.011417l^2 & 0.203539l & -0.007479l^2 \\ 0.041701l & 0.081199l & \ldots & \ldots & \ldots \\ -0.520183 & \ldots & 0.043182l & \ldots & \ldots \\ 0.092145l & \ldots & \ldots & 0.112375 & \ldots \\ -0.961140 & \ldots & \ldots & \ldots & 0.030427l \end{bmatrix}^*$$

*Note: The product $K^{-1}M$ of 2 symmetric matrices is not symmetric (nearly always true!) —General Editor.

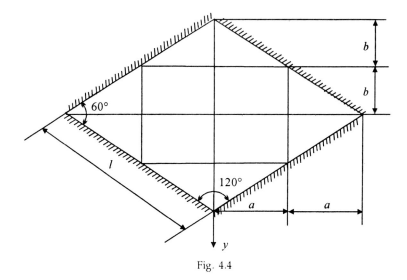

Fig. 4.4

When reproducing this matrix in the cited work, its values which were not used in the calculations using eqns. (4.50), (4.52) and (4.55) were omitted. Having selected a value in the left upper corner of matrix C as the centre of reduction (condensation point), we find from eqns. (4.54) and (4.52) that

$$\lambda_{max} = 0.42978 \frac{\overline{m_0} l^4}{D}, \quad \overline{m_0} = m_0 \cdot 10^3,$$

$$\text{or } \omega_{min} = 1.5254 \sqrt{\frac{1000 D}{m_0 l^4}}.$$

This value differs from $\omega_{min} = 1.532766 \sqrt{\dfrac{1000 D}{m_0 l^4}}$ given in the work cited above by 0.48%.

Considering that when condensing the masses of each of the 't' groups of auxiliary degrees of freedom to nodes 'r', the maximum eigenvalue for the partial system with a small number of degrees of freedom has to be determined, it would be most convenient to use in this case the iteration method (Shoop, 1982; Chapter 2) best suited for determining the maximum and minimum eigenvalues. These methods are least sensitive to rounding errors although they help determine not more than three frequencies fairly accurately.

When using iteration methods, the rapidity of convergence of the solutions depends on the selection of the initial vector. If it is close to the true eigenvector, the process converges very rapidly.

Therefore, eqn. (4.54) can be used for finding the λ_{max} value of a partial system. From the maximum eigenvalue, the eigenvector corresponding to it can be found and later rendered accurately by the iteration method.

The above algorithm was incorporated in the DINKON program developed by Blokhina (1989; Chapter 2).

Example 4.5. A square isotropic plate rigidly fixed along the contour was analysed (Fig. 4.5) (Blokhina, 1989; Chapter 2). A 6 × 6 grid of nodes laid on the plate divided it into 36 FE. The Poisson coefficient for the plate material has been taken as zero.

Nodes '*r*' depicted as black dots in Fig. 4.5 have been adopted as the main nodes (master dofs) nodes. All the slave degrees of freedoms have been divided into three groups S_1, S_2, S_2. Considering that the mass of the plate was reduced to the nodes of the FE grid and that, during the oscillations of the reduced point loads, inertial forces arise only along the direction of their linear displacements from the plane of the plate, non-inertial degrees of freedom of the system should be excluded from the analysis when using the finite element method. This is done by using the renumeration matrix obtained by transforming the identity matrix (see eqn. (1.9)).

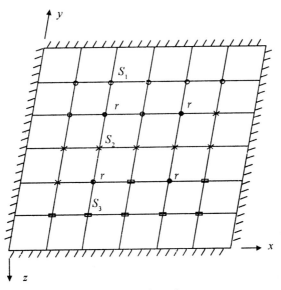

●	master (main) nodes
○	1-group of slave dofs
×	2-group of slave dofs
▢	3-group of slave dofs

Fig. 4.5

Using the renumeration matrix, let us convert the stiffness and mass matrices, whereby the elements of these matrices relating to inertial and non-inertial degrees of freedom are grouped into blocks:

$$K_1 = C^T KC = \begin{bmatrix} K_{\varphi\varphi} & K_{z\varphi} \\ K_{\varphi z} & K_{zz} \end{bmatrix},$$ ... (4.56)

$$M_1 = C^T MC = \begin{bmatrix} 0 & 0 \\ 0 & M_{zz} \end{bmatrix}.$$ ... (4.57)

The above conversion helps eliminate in the following equation

$$|K - \lambda M| = 0,$$ ... (4.58)

the block of unknown rotation angles according to Gauss and to derive the equation

$$|M_{zz}^{-1} \tilde{K}_{zz} - \lambda I| = 0_1,$$ ... (4.59)

in which M_{zz} is the diagonal matrix of mass; and $\tilde{K}_{zz} = K_{zz} - K_{\varphi z} K_{\varphi\varphi}^{-1} K_{z\varphi}$ the reduced matrix of stiffness.

By solving the system equations (4.59), we obtain the natural frequency values for the plate considered for the initial mesh of FE nodes.

Table 4.3 provides a comparison of these results with the accurate results given in *Dynamic Analysis of Buildings and Structures* (1984; Chapter 2) and also a comparison of the results obtained by the proposed method of dynamic condensation with the solutions for the original unreduced problem.

Table 4.3

ω_i	Accurate solution	Finite element method, 6 × 6	Δ, %, of accurate solution	Dynamic condensation ($n = 4$)	Δ, %, of finite element method	Δ, %, of accurate solution
ω_1	33.99	35.3	1.9	35.06	0.67	2.58
ω_2	73.41	71.73	2.2	70.89	1.17	3.43
ω_3	73.41	71.73	2.2	70.89	1.17	3.43
ω_4	108.27	102.86	4.9	101.07	1.74	6.65
ω_5	131.64	129.35	1.7	-	-	-
ω_6	132.25	130.44	1.3	-	-	-
ω_7	165.15	156.27	5.3	-	-	-
ω_8	165.15	156.27	5.3	-	-	-
ω_9	210.50	204.34	2.9	-	-	-
ω_{10}	210.50	204.34	2.9	-	-	-

To understand the effect of the number of condensation nodes on the accuracy of determining the eigenvalues, mass condensation was carried out consecutively to one, five, nine and thirteen nodes (Fig. 4.6). The results of comparison with the frequency values of the finite element method (Table 4.3) are given in Table 4.4.

Table 4.4

r	ω_i	$\Delta, \%$	$\omega_2 = \omega_3$	$\Delta, \%$	ω_4	$\Delta, \%$
1	35.520	1.82	-	-	-	-
5	35.065	0.66	71.01	1.0	101.09	1.72
9	35.200	0.28	71.45	0.39	102.11	0.73
13	35.360	0.16	71.79	0.25	103.52	0.64

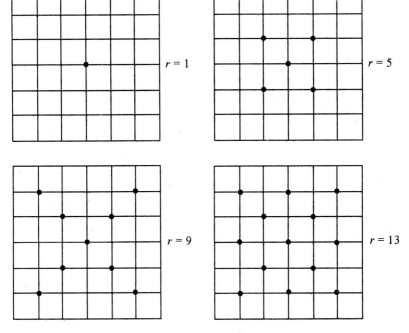

Fig. 4.6

It can be seen from the above results that use of the proposed method of consecutive dynamic condensation ensures the required accuracy of calculating the frequencies of natural oscillations. Special mention should be made of machine resource economy achieved in this method.

4.8 Consecutive Frequency-dynamic Condensation Using Reduced Spectrum of Eigenvalues and Eigenvectors of Subsystems

Let us rewrite eqn. (4.39) in the form

$$\left[D_{rr}^{(r+s)} (\lambda) \right] \{z_r\} = \{z_r\} \lambda, \qquad \qquad \dots (4.60)$$

in which $\left[D_{rr}^{(r+s)} \left(\lambda \right) \right] = \delta_{rr} M_r + D_{rr}^{(s)} \left(\lambda \right).$ ... (4.61)

Expression (4.60) is true to all the r eigenvalues λ_k of the original eqn. (4.1) for the i-th partial system:

$$\left[D_{rr}^{(r+s_i)} \left(\lambda \right) \right] \left[z_{rk}^{(i)} \right] = \left[z_{rk}^{(i)} \right] \left[\lambda_k^{(i)} \right],$$ ... (4.62)

in which $\left[D_{rr}^{(r+s_i)} \right]$ is the approximating matrix of coefficients during the condensation of the i-th partial system to the main degrees of freedom; $\left[z_{rk}^{(i)} \right] = \left[\bar{z}_{r1}^{(i)}, \bar{z}_{r2}^{(i)}, \ldots, \bar{z}_{rr}^{(i)} \right]$ the matrix of eigenvectors of the i-th partial system and

$$\left[\lambda_k^{(i)} \right] = \begin{bmatrix} \lambda_1^{(i)} & & & \\ & \lambda_2^{(i)} & & \\ & & & \\ & & & \\ & & & \lambda_r^{(i)} \end{bmatrix}$$

is the diagonal matrix of eigenvectors of the i-th partial system.

The approximating matrix of coefficients $\left[D_{rr}^{(r+s_i)} \right]$ is determined on the basis of equilibrium (4.62) using the equation

$$\left[D_{rr}^{(r+s_i)} \right] = \left[z_{rk}^{(i)} \right] \left[\lambda_k^{(i)} \right] \left[z_{rk}^{(i)} \right]^{-1}.$$ ... (4.63)

From eqn. (4.61), we find the matrix of dynamic conversion, i.e., the matrix of condensation additives:

$$\left[D_{rr}^{(s)_i} \left(\lambda \right) \right] = \left[D_{rr}^{(r+s_i)} \right] - \left[\delta_{rr} M_r \right] = \Delta^{(i)} \left[\delta_{rr} M_r \right].$$ ... (4.64)

Summing up these additives for all the t partial systems, we obtain:

$$\left[D_{rr}^{(r+s)} \right] = \delta_{rr} M_r + \sum_{i=1}^{t} \Delta^{(i)} \left[\delta_{rr} M_r \right].$$ ... (4.65)

in which t is the number of partial systems (groups of slave degrees of freedom).

Having solved the characteristic equation

$$\left[D_{rr}^{(r+s)} - \lambda E_r \right] = 0, \qquad \qquad \text{... (4.66)}$$

we find r eigenvalues and eigenvectors corresponding to them for the system condensed to the main degrees of freedom. The missing components of eigenvectors can be restored when required using eqn. (4.38).

The accuracy of the two variants of the consecutive frequency-dynamic condensation method has been analysed comparatively on the example of equal placed point masses for a beam hinged at the ends; $m_i = m = 1$ (Fig. 4.7).

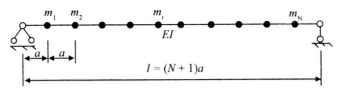

Fig. 4.7

Table 4.5 gives

a) accurate values of natural frequencies (column 4) for the case $N = 12$ calculated using eqn. (2.5.1) of Ignatiev (1979, p. 92):

$$\omega_k^2 = \frac{48\,EJ}{ma^3} \cdot \frac{\sin^4 \dfrac{k\pi}{2(N+1)}}{2 + \cos \dfrac{k\pi}{N+1}};$$

b) natural frequency values obtained by the method of static condensation (column 5) for different numbers of the primary degrees of freedom n (after conversion of the flexibility matrix and change over to equations of the displacement method); and

c) natural frequency values obtained by the first (column 6) and second variants (column 7) of consecutive frequency-dynamic condensation method for different numbers of the main degrees of freedom n and different numbers of slave degrees of freedom s_i in the partial system.

Table 4.6 gives the results of analysis for $N = 104$ carried out by the second variant of consecutive frequency-dynamic condensation and compared with the accurate solution. As in the preceding example, the effect of the number of slave degrees of freedom s_i in the partial system on the accuracy of the result was studied.

Table 4.5

Scheme of arrangement of condensation nodes	s_i	Number of frequencies, k	Accurate solution, ω_k	Static condensation	1st variant of frequency-dynamic condensation	2nd variant of frequency-dynamic condensation
					frequency ω_k (Δ,%)	
1	2	3	4	5	6	7
I	1	1	0.13328	0.13183 (1.1)	0.13950 (4.6)	· 0.13525 (−1.4)
II	1	1	0.13328	0.13339 (0.1)	0.13752 (3.1)	0.13375 (−0.4)
	2				0.1387 (4.0)	0.13383 (−0.4)
	5				0.13903 (4.3)	0.13349 (−0.2)
	1	2	0.00833	0.01112 (33.0)	0.0115 (38.0)	0.00851 (−2.2)
	2				0.0116 (39.1)	0.0085 (2.0)
	5				0.01162 (39.5)	0.00855 (−2.6)
III	1	1	0.13328	0.1349 (1.2)	0.14354 (7.7)	0.13351 (−0.2)
	2				0.14388 (7.9)	0.13345 (−0.13)
	4				0.14417 (8.1)	0.13345 (−0.1)
	1	2	0.00833	0.00818 (1.8)	0.00785 (5.7)	0.00836 (−0.4)
	2				0.00787 (5.5)	0.00834 (−0.1)
	4				0.00789 (5.2)	0.00836 (−0.4)
	1	3	0.00165	0.00177 (7.2)	0.00156 (5.4)	0.00162 (1.8)
	2				0.00156 (5.4)	0.00162 (1.8)
	4				0.00158 (4.2)	0.00163 (1.2)
	1	4	0.00052	0.000898 (72.7)	0.00087 (67.3)	0.00054 (−5.7)
	2				0.00087 (67.3)	0.00052 (0.0)
	4				0.00087 (67.3)	0.00053 (−1.9)
IV	1	1	0.13328	0.1346 (1.0)	0.14277 (7.1)	0.13346 (−0.1)
	2				0.14355 (7.8)	0.13347 (−0.1)
	3				0.14329 (7.5)	0.13345 (−0.1)
	1	2	0 00833	0.00884	0.0096 (15.6)	0.00834 (−0.1)
	2				0.00967	0.00834

Contd.

Table 4.5 (contd.)

1	2	3	4	5	6	7
	3			(6.1)	(16.1) 0.00965 (16.1)	(-0.1) 0.00834 (-0.1)
IV	1				0.00143 (13.3)	0.00162 (1.8)
	2	3	0.00165	0.00139 (15.7)	0.00144 (12.7)	0.00165 (0.0)
	3				0.00144 (12.7)	0.00164 (0.1)
	1				0.00063 (21.1)	0.00053 (-1.9)
	2	4	0.00052	0.00065 (25.0)	0.00064 (23.1)	0.00053 (-1.9)
	3				0.00063 (21.1)	0.00052 (0.0)
	1				0.00017 (19.0)	0.00023 (-9.5)
	2	5	0.00021	0.00019 (9.5)	0.00017 (19.0)	0.00019 (9.5)
	3				0.00017 (19.0)	0.00020 (4.7)
	1				0.00016 (60.0)	0.00012 (-20.0)
	2	6	0.00010	0.00017 (70.7)	0.00016 (60.0)	0.00010 (0.0)
	3				0.00016 (60.0)	0.00011 (9.5)

The schemes showing the arrangement of condensation nodes are depicted below:

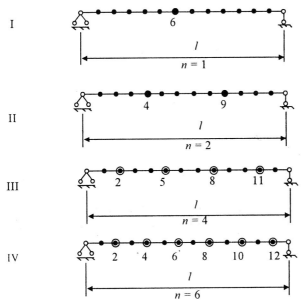

Table 4.6

N, n	Number of frequency	s_i	Accurate solution, ω_k	Second variant of frequency-dynamic condensation	
				frequency ω_k	Δ, %
N = 104 n = 4	1	1 2 5	1.0779	1.07825 1.07856 1.07880	0.03 0.06 0.08
	2	1 2 5	0.06737	0.06765 0.06783 0.06806	0.3 0.6 1.0
	3	1 2 5	0.01331	0.01323 0.01347 0.01370	0.6 1.2 3.0
	4	1 2 5	0.00421	0.00430 0.00443 0.00453	2.1 5.2 7.6

Table 4.7 shows the results characterising the effect of values s_i on machine time consumption.

Table 4.7

N	n	s_i	Machine time consumption
N = 12	4	1 2 4	1'32" 1'31" 1'33"
	6	1 2 3	1'48" 1'41" 1'38"
N = 104	4	1 2 5	3'26" 2'57" 3'10"

The following conclusions pertaining to the consecutive frequency-dynamic condensation method can be drawn on the basis of comparative analysis:

(1) on increasing the number of the main degrees of freedom n, the accuracy of calculating the eigenvalues increases and

(2) the method helps obtain eigenvalues very close to the accurate values for the entire reduced spectrum but not only for the maximum eigenvalue (as in the method of static condensation).

The number of groups into which the slave degrees of freedom are divided does not significantly influence the accuracy of calculating the eigenvalues and eigenvectors. It has little influence on machine time consumption.

A square plate with uniformly distributed mass $m = 1$ and different variants of supporting the contour sides was selected as another example for comparative analysis. Analysis of such a plate by the finite element method has been discussed in *Structural Analysis* by numerical methods. (Maslennivov, 1987; Chapter 2) in which the values of the stiffness matrix coefficients and equivalent masses have been given for the adopted rectangular FE with 12 degrees of freedom. The flexibility matrices required for analysis by the method of consecutive frequency-dynamic condensation were obtained by inverting the corresponding stiffness matrices.

The results of analyses are shown in Tables 4.8, 4.9 and 4.10. Table 4.8 shows the results of calculations for different values of s_i at constant number of the main degrees of freedom (condensation nodes) n. s_i is the number of slave dof's in each condensed block. Table 4.9 compares the results for the two variants of condensation nodes, their total number remaining identical. Table 4.10 gives the results of a numerical study of the convergence of solutions on increasing the number of condensation nodes.

Table 4.8

| FEM scheme of plate | s_i | Number of frequencies, k | Frequency ω_k values calculated by | | |
			finite element method	static condens- ation method	block-wise frequency dynamic condensation
1	2	3	4	5	6
	1				0.1479 (−0.5)
	2	1	0.1487	0.141	0.1484 (−0.2)
	11			(−5)	0.1487 (0.0)
	1				0.0363 (−3.0)
	2	2	0.03762	0.031	0.0369 (−1.8)
	11			(−17)	0.03752 (−0.3)
$N = 28$	1				0.0167 (−4.0)
$n = 6$	2	3	0.01747	0.0161	0.0171 (−1.7)
	11			(−7)	0.01729 (−1.0)
	1				0.0105 (−4.0)
	2	4	0.01107	0.009	0.0106 (−3.0)
	11			(−18)	0.01095 (−1.1)
	1				0.00946 (−3.2)
	2	5	0.00977	0.0081	0.00960 (−1.7)
	11			(−17)	0.00969 (−8.8)
	1				0.06459 (−4.0)
	2	6	0.006727	0.0053	0.06520 (−3.0)
	11			(−21)	0.06680 (−0.7)

Table 4.9

FEM scheme of plate	s_i	k	finite element method	static condensation	Δ%	block-wise frequency dynamic condensation	Δ%
1	2	3	4	5		6	
$N = 39\ n = 5$	1	1	0.0950	0.0934	(−1.7)	0.09435	(−0.7)
		2	0.0228	0.0189	(−17)	0.0219	(−3.9)
		3	0.0228	0.0189	(−17)	0.0219	(−3.9)
		4	0.0116	–		0.011507	(−0.9)
		5	0.0069	0.00636	(−8.1)	0.0069	(0)
	2	1	0.0950	0.0934	(−1.7)	0.0929	(−0.55)
		2	0.0228	0.0189	(−17)	0.2210	(−3.2)
		3	0.0228	0.0189	(−17)	0.2210	(−3.2)
		4	0.0116	–		0.01152	(−0.7)
		5	0.0069	0.00636	(−8.1)	0.006885	(−0.7)
$N = 39$ $n = 5$	3	1	0.0950	0.0934	(−1.7)	0.0940	(−1.0)
		2	0.0228	0.0189	(−17)	0.02199	(−4.0)
		3	0.0228	0.0189	(−17)	0.02199	(−4.0)
		4	0.0116	–		0.0113	(−2.5)
		5	0.0069	0.00636	(−8.1)	0.00686	(−0.6)

Table 4.10

Analytical scheme of plate	s_i	k	finite element method	static condensation	block-wise frequency dynamic condensation	
1	2	3	4	5	6	
$N = 36,\ n = 1$	1	1	0.1947	0.1677 (−13.0)	0.1794	(−8.0)
	2				0.1823	(−6.0)
	3				0.1860	(−4.5)
	5				0.1947	(0.0)
$N = 36,\ n = 2$	1	1	0.1947	0.1676 (−14.0)	0.1823	(−6.0)
	2				0.1859	(−4.5)
	4				0.1887	(−3.0)
	8				0.1927	(−1.0)
	11				0.1947	(−0.0)
	1	2	0.0756	0.0651 (−13.0)	0.0692	(−8)
	2				0.0680	(−6.5)
	4				0.0717	(−5)
	8				0.0745	(−1)
	11				0.0755	(0)
$N = 36,\ n = 3$	1	1	0.1947	0.1731 (−11.0)	0.1831	(−5.8)
	3				0.1861	(−4.3)
	8				0.1906	(−2.1)
	11				0.1942	(−0.2)
	1	2	0.0756	0.0444 (−41.0)	0.0694	(−8.0)
	3				0.0712	(−5.8)

(Contd.)

Table 4.10 (Contd.)

FEM scheme of plate	s_i	Number of frequencies, k	Frequency ω_k calculated by		block-wise frequency dynamic condensation
			finite element method	static conden-sation	block-wise frequency dynamic condensation
	8				0.0733 (-3.0)
	11				0.0754 (-0.3)
	1				0.0239 (-14)
	3	3	0.0277	0.0192	0.0245 (-11.0)
	8			(-30.0)	0.0255 (-7.0)
	11				0.0267 (-7)
	1				0.1759 (-9.0)
	2	1	0.1947	0.1636	0.1794 (-7.5)
	4			(-16.0)	0.1864 (-4)
	8				0.1919 (-1.5)
	16				0.1933 (-0.3)
	1				0.0705 (-7.0)
	2				0.0711 (-5.9)
$N = 36, n = 4$	4	2	0.0756	0.523	0.0728 (-3.7)
	8			(-30.0)	0.0733 (-3.0)
	16				0.0744 (-1.5)
	1				0.0258 (-6.8)
	2				0.0261 (-5.0)
	4	3	0.0277	0.0157	0.0268 (-3.9)
	8			(-43.0)	0.0271 (-2.0)
	16				0.0274 (-0.8)
	1				0.01702 (-18)
	2				0.0179 (-13)
	4	4	0.0206	0.0136	0.01966 (-5.0)
	8			(-34.0)	0.02010 (-2.5)
	16				0.02043 (-0.9)
	1				0.1872 (-3.8)
	2				0.1874 (-3.8)
	3	1	0.1947	0.1778	0.1899 (-2.4)
	10			(-8.0)	0.1932 (-0.8)
	1				0.0725 (-4.0)
	2				0.0733 (-3.0)
	3	2	0.0756	0.0563	0.0740 (-2.0)
	10			(-25.0)	0.0750 (-0.9)
$N = 36,$	1				0.0259 (-6.4)
$n = 6$	2				0.0269 (-2.8)
	3	3	0.0277	0.0222	0.0271 (-2.7)
	10			(-20.0)	0.0275 (-0.6)
	1				0.0184 (-11.0)
	2				0.0189 (-8.2)
	3	4	0.0206	0.0106	0.0192 (-7.0)
	10			(-48.0)	0.0198 (-3.8)
	1				0.01808 (-6)
	2				0.01818 (-5)
	3	5	0.0192	0.0171	0.01870 (-2.5)
	10			(-11.0)	0.0191 (-0.5)
	1				0.0087 (-10.0)
	2				0.0089 (-8.0)
	3	6	0.0097	0.0078	0.0090 (-7.0)
	10			(-20.0)	0.0094 (-3.0)

4.9 Consecutive Frequency-dynamic Condensation when Analysing by Displacement or Finite Element Method

Based on eqn. (4.35), the following algorithm can be constructed for solving the partial problem of eigenvalues and eigenvectors of eqn. (4.6).

As in Section 4.7, let us divide the auxiliary unknowns into t groups. The characteristic matrix for eqn. (4.6) corresponding to this division of unknowns into groups has the following form:

$$K - \lambda M = \begin{vmatrix} \boxed{d_{11}} & d_{12} & \cdots & \boxed{d_{1r}} & \cdots & d_{1s} & \cdots & d_{1m} \\ d_{21} & \boxed{d_{22}} & \cdots & \boxed{d_{2r}} & \cdots & d_{2s} & \cdots & d_{2m} \\ \cdots & \cdots & \cdots & \cdots & \cdots & \cdots & \cdots & \cdots \\ \boxed{d_{r1}} & \boxed{d_{r2}} & \cdots & \boxed{d_{rr}} & \cdots & \boxed{d_{rs}} & \cdots & \boxed{d_{rm}} \\ \cdots & \cdots & \cdots & \cdots & \cdots & \cdots & \cdots & \cdots \\ d_{s1} & d_{s2} & \cdots & \boxed{d_{sr}} & \cdots & \boxed{d_{ss}} & \cdots & d_{sm} \\ \cdots & \cdots & \cdots & \cdots & \cdots & \cdots & \cdots & \cdots \\ d_{m1} & d_{m2} & \cdots & \boxed{d_{mr}} & \cdots & d_{ms} & \cdots & \boxed{d_{mm}} \end{vmatrix} = 0, \quad \dots (4.67)$$

in which, in a general case:

$$d_{rr} = K_{rr} - \lambda M_{rr}, \, d_{sr} = K_{sr} - \lambda M_{sr}.$$

Using eqns. (4.7) and (4.35), the secondary (slave) unknowns are eliminated and the stiffness and mass matrices condensed blockwise. The results are substituted in eqn. (4.31).

If the same condensation nodes are used in each stage of blockwise condensation, the characteristic equation for the condensed system assumes the following form:

$$\left| (K_{rr} - \lambda M_{rr}) + \sum_{S=1}^{t} K_{rr}^{(s)} - \lambda \sum_{S=1}^{t} M_{rr}^{(s)} \right| = 0. \qquad \dots (4.68)$$

In many cases, the use of the same condensation nodes in all stages is not possible because of the wide dispersal of coefficient submatrices in the system of equations

$$|K - \lambda M| = 0$$

and an absence of coupling between some blocks of the secondary unknowns and the block of the primary unknowns. In this case, the use of 'sliding' condensation nodes similar to the use of 'sliding' selected points in substructuring analysis is most effective (Sapozhnikov, 1980a, 1980b; Chapter 1).

For the first stage of condensation, the characteristic equation can be expressed as:

$$\left| \left(K_{rr}^{(i)} - \lambda M_{rr}^{(i)} \right) + \left(K_{rr}^{(s),i-1} - \lambda M_{rr}^{(s),i-1} \right) \right| = 0. \qquad \dots (4.69)$$

The values of $\lambda^{(i)}_{1,r,s}$ and $\lambda^{(i)}_{2,r,s}$ found by solving eqn. (4.69) are used in the equations for $M^{(s),i}_{r,r}$ and $K^{(s),i}_{r,r}$ and in the characteristic eqn. (4.69) at the $(i + 1)$-th stage.

To evaluate the accuracy of the algorithm described by Blokhina (1989; Chapter 2), the plate depicted in Figure 4.8 (example 4.5, Section 4.7) has been analysed again. The number of condensation nodes selected was nine, selected in such a way that the form of oscillations could be restored using the spline interpolation method described in Section 2.3.

Solving eqn. (4.68) gave nine frequencies of the natural oscillations of the plate whose values are given in Table 4.11.

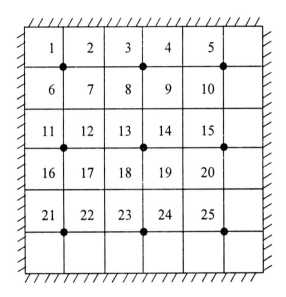

Fig. 4.8

Table 4.11

ω	Accurate solution	Finite element method, 6 × 6	Dynamic condensation	Δ, %, of accurate solution	Δ, %, of finite element method
ω_1	35.99	35.3	35.24	2.08	0.17
ω_2	73.41	71.73	71.49	2.62	0.33
ω_3	73.41	71.73	71.49	2.62	0.33
ω_4	108.27	102.86	103.76	4.16	0.88
ω_5	131.64	129.35	127.96	2.8	1.07
ω_6	132.25	130.44	128.04	3.18	1.84
ω_7	165.15	156.27	151.79	8.08	2.86
ω_8	165.15	156.27	151.79	8.08	2.86
ω_9	210.5	204.34	198.15	5.86	3.09

The amplitude values of the first and ninth forms of the nodal oscillations in the original grid calculated by the finite element method are compared in Table 4.12 with the displacements of the same nodes calculated by the suggested algorithm using spline interpolation. This comparison reveals the good accuracy of the results obtained by the proposed method.

Table 4.12

z	First form			Ninth form		
	finite element method	dynamic condensation $r = 9$	Δ, %	finite element method	dynamic condensation $r = 9$	Δ, %
z_1	0.089156	0.9001	0.96	−0.92995	−0.9085	2.3
z_2	0.24266	0.24003	1.08	−0.17893	−0.1703	4.8
z_3	0.31137	0.30004	3.63	−0.98859	0.9239	6.5
z_4	0.24266	0.24004	1.07	−0.17895	−0.1704	4.7
z_5	0.08916	0.09001	0.95	−0.95998	−0.9391	2.2
z_6	0.24266	0.23585	2.80	−0.17888	−0.1707	4.5
z_7	0.62292	0.62895	0.97	−0.04458	−0.04291	3.7
z_8	0.79041	0.78624	0.53	0.178	0.1691	5.2
z_9	0.62292	0.62904	0.98	−0.04457	−0.04308	3.3
z_{10}	0.24266	0.23589	2.78	−0.17892	−0.1697	5.1
z_{11}	0.31137	0.29912	3.93	0.98866	0.9499	3.9
z_{12}	0.79041	0.7977	0.92	0.17798	0.17352	2.5
z_{13}	1.0000	0.99727	0.27	−1.000	0.9698	3.0
z_{14}	0.79041	0.7979	0.95	0.17802	0.17521	0.3
z_{15}	0.31137	0.29924	3.89	0.98864	0.9492	3.9

Blokhina (1989; Chapter 2) incorporated the proposed method of consecutive frequency-dynamic condensation in the form of the displacement method in a program package. Two examples of the calculations given below were worked out by Blokhina to illustrate the possibilities of this method.

Example 4.6. A thin-walled cellular structure depicted in Figure 4.9 has the following parameters: $\delta = 0.05$ m, $v = 0.3$ and $E = 2.1 \times 10^5$ MPa. The geometric dimensions of the structure are shown in the Figure.

Taking into consideration the spatial behaviour of the structure in each node of the finite element grid, six degrees of freedom were selected. Condensation nodes r depicted as dots in the Figure were selected such that the forms of oscillations could be plotted by spline approximation using the amplitudes of their displacements.

The inner nodes s have been divided into four groups conforming to the four planes constituting the cellular shell and condensed to the main nodes r.

By solving the condensed characteristic equation, 24 frequencies and the forms of oscillations corresponding to them were determined (Table 4.13). The results are compared with the solution in the initial FE mesh.

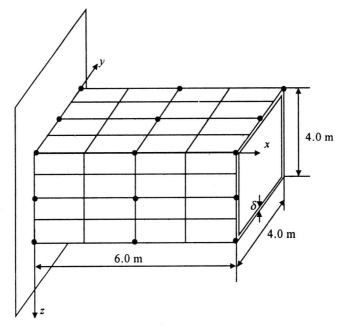

Fig. 4.9

Table 4.13

ω	Finite element method	Dynamic condensation ($n = 32$)	Δ, %
ω_1	10.51	10.22	2.76
ω_2	15.035	14.61	2.77
ω_3	15.035	14.61	2.77
ω_4	15.774	15.01	4.84
ω_5	18.273	17.73	2.97
ω_6	18.273	17.73	2.97
ω_7	19.267	18.12	5.95
ω_8	21.087	20.43	3.11

Example 4.7. This example covers a rectangular bevelled cellular shell (Fig. 4.10) whose analytical results have been given by Blokhina and Ignat'eva (1987; Chapter 2).

The analysis was carried out for the following values of parameters: $\alpha = 0$ and $\alpha = 45°$, $\nu = 0.3$, $\delta = 0.005\ l$ and $d/l = 0.5$.

The results of analysing by the proposed method are compared in Table 4.14 with the results reported by Craig and Bampton (1968; Chapter 2).

The results of analysing thin-walled cellular shells based on the proposed consecutive dynamic condensation method revealed its high efficiency in terms of reduced computer time consumption while providing the possibility of obtaining results with any required degree of accuracy.

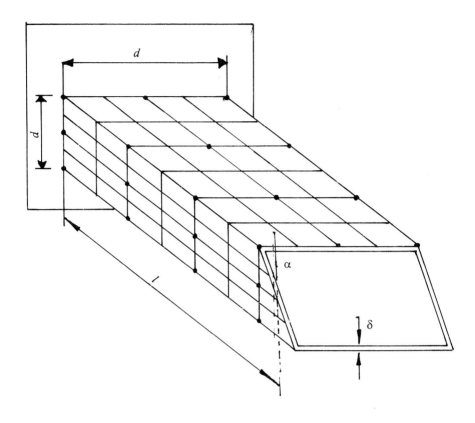

Fig. 4.10

Table 4.14

ω	α = 0°			α = 45°		
	according to Craig (1968; Chapter 2)	Dynamic condensation	Δ, %	according to Craig (1968; Chapter 2)	Dynamic condensation	Δ, %
ω_1	0.0606	0.0593	2.14	0.0583	0.0569	2.4
ω_2	0.09120	0.0865	5.17	0.079	0.0752	4.81
ω_3	0.09121	0.0865	5.17	0.0866	0.0921	6.35
ω_4	0.1166	0.1097	5.9	0.0998	0.1019	2.1
ω_5	0.1312	0.1256	4.26	0.1112	0.1157	4.05
ω_6	0.1506	0.1473	2.19	0.119	0.1238	4.03
ω_7	0.1509	0.1473	2.19	0.131	0.1391	6.18
ω_8	0.1717	0.1601	6.76	0.1401	0.1522	6.83

4.10 Improvement of Eigenvalues and Eigenvectors by Iteration

For solving several dynamic problems, a knowledge not only of eigenvalues but also their corresponding eigenvectors is required. Finding the eigenvectors of the system of eqns. (4.1) for the known eigenvalues λ_k ($k = 1, 2,..., n$) is straight forward and involves solving a system of $(n - 1)$ linear algebraic equations obtained by eliminating any of the equations in system (4.1).

Adopting, for example, in eqn. (4.1) $z_{1k} = 1$ and transferring into the right-hand parts the resultant free members, we obtain the following system of equations:

$$c_{12} z_2 + c_{13} z_3 + \cdots + c_{1n} z_n = \left(\lambda_k - c_{11}\right),$$

$$\left(c_{22} - \lambda_k\right) z_2 + c_{23} z_3 + \cdots c_{2n} z_n = -c_{21},$$

$$c_{32} z_2 + \left(c_{33} - \lambda_k\right) z_3 + \cdots + c_{3n} z_n = -c_{31}, \qquad \ldots (4.70)$$

$$\cdots\cdots\cdots\cdots\cdots\cdots\cdots\cdots\cdots\cdots\cdots$$

$$c_{n2} z_2 + c_{n3} z_3 + \ldots + \left(c_{nn} - \lambda_k\right) z_n = -c_{n1}.$$

By isolating in particular the first equation of the system, the remaining $(n - 1)$ equations are solved for known λ_k and $(n - 1)$ components of vector $\{z_k\}$, i.e., $z_{2k}, z_{3k}, \ldots, z_{nk}$, are found. When required, the resultant eigenvectors are normalised.

The obtained values of vector component $\{z_k\}$ can be substituted in the first equation to verify the accuracy of the solution and also to build up the iteration process for improving the value of λ_k and its corresponding vector $\{z_k\}$.

The iteration process is as follows.

The eigenvalue λ_k found by an approximate method, e.g., by one of those discussed above, is introduced in eqn. (4.70) and solved relative to vector component $\{z_k\}$. This gives the zero approximation vector:

$$\{z_k^0\} = \left\lfloor 1, z_{2k}^{(0)}, z_{3k}^{(0)}, \ldots, z_{Nk}^{(0)} \right\rfloor^T$$

Stepwise improvement proceeds thereafter. By substituting vector $\{z_k^{(D)}\}$ in equation

$$[c] \{z\} = \lambda \{z\}, \qquad \ldots (4.71)$$

we obtain

$$[c]\{z_k^{(0)}\} = \left\lfloor z_{1k}^n, z_{2k}^n, \ldots, z_{Nk}^n \right\rfloor^T = \lambda_k^{(1)} \left\lfloor 1, z_{2k}^{(1)}, \ldots, z_{Nk}^{(1)} \right\rfloor^T = \lambda_k^{(1)} \{z_k^{(1)}\}. \qquad \ldots (4.72)$$

In the second step, vector $\{z_k^{(1)}\}$ is introduced in eqn. (4.71) and the improved value $\lambda_k^{(2)}$ and improved vector component $\{z_k^{(2)}\}$ are found.

The solution can be organised for any degree of accuracy. The following difference can serve to control accuracy:

$$\frac{1}{\lambda_k^{(i)}}\left[\lambda_k^{(i+1)} - \lambda_k^{(i)}\right] \leq \varepsilon \qquad \qquad \text{... (4.73)}$$

in which ε is the accuracy of calculations and i the number of the iteration.

The convergence and accuracy of this algorithm were evaluated for the example analysing the plate studied in Section 4.7 (Example 4.5). Condensation was done to four nodes ($n = 4$) as shown in Figure 4.5. Improved values of λ_k as well as amplitudes of oscillation forms (eigenvectors) corresponding to the first frequency were obtained by applying the iteration procedure to improve the eigenvalues.

The comparison (Table 4.15) carried out with the results obtained before by the finite element method (see Tables 4.3, 4.11 and 4.12) verifies the rapid convergence of the algorithm and the high accuracy of results.

Table 4.15

ω	$n = 4$ Improved value	Δ, %	z	First form	Δ, %
ω_1	35.25	0.14	z_7	0.6312	1.33
ω_2	71.36	0.52	z_9	0.6309	1.28
ω_3	71.36	0.52			
ω_4	102.01	0.83			

It can be seen from a comparison of Tables 4.15 and 4.11 that the error in determining the frequencies after improvement by iteration (for $n = 4$) does not exceed the error of condensation to nine nodes ($n = 9$) without following the iteration procedure.

4.11 Consecutive Frequency-dynamic Condensation as a Displacement Method Using the Full Spectrum of Eigenvalues and Eigenvectors of Partial Systems

Let us examine the problem of reducing the number of degrees of freedom of the subsystem described by equations of the form of (4.6) or (4.7) from the viewpoint of the equivalence of the reduced subsystem of the initial trial with respect to eigenvalues and the corresponding eigenvector components.

The following equation is available for a subsystem reduced to the main degrees of freedom:

$$\left[\tilde{K}_{rr} - \lambda \tilde{M}_{rr}\right]\{z_r\} = 0, \quad r = 1, 2, \ldots, n, \qquad \text{... (4.74)}$$

in which $\tilde{K}_{rr}$ is the stiffness matrix of the subsystem condensed to the printing degrees of freedom and $\tilde{M}_{rr}$ the condensed mass matrix of the subsystem.

Since the eigenvalues of this equation and the eigenvector components should coincide with the first of the n eigenvalues and the corresponding components of the first n eigenvectors, the following equation can be expressed:

$$\left[\tilde{K}_{rr}\right]\left[z_{rk}\right] = \left[\tilde{M}_{rr}\right]\left[z_{rk}\right]\left[\lambda_k\right],$$ ... (4.75)

$$k = 1, 2, \ldots, n$$

in which $\left[z_{rk}\right] = \left[\tilde{z}_{1k}, \tilde{z}_{2k}, \ldots, \tilde{z}_{rk}, \ldots, \tilde{z}_{nk}\right]$ is a quadratic matrix of order $n \times n$ whose columns represent the first n eigenvectors of eqn. (4.7) with components pertaining only to the primary degrees of freedom r; and λ_k is a diagonal matrix of the first n eigenvalues of the subsystem.

Equation (4.75) can be solved relative to $\tilde{M}_{rr}$ or $\tilde{K}_{rr}$.

From the physical concept, the first variant is more appropriate. In it, stiffness matrix $\tilde{K}_{rr}^{*}$ which was statically condensed using eqn. (4.12), can be adopted as matrix $\tilde{K}_{rr}$ while the condensed mass matrix can be represented as:

$$\tilde{M}_{rr} = M_{rr}^{*} + \Delta M_{rr},$$ ... (4.76)

in which M_{rr}^{*} is the mass matrix statically condensed using eqn. (4.12) and ΔM_{rr} the matrix of condensed masses.

Equation (4.75) can now be expressed as:

$$\left[K_{rr}^{*}\right]\left[z_{rk}\right] = \left[M_{rr}^{*} + \Delta M_{rr}\right]\left[z_{rk}\right]\left[\lambda_k\right].$$ ... (4.77)

On solving eqn. (4.77) relative to the condensed mass matrix, the following equations are found:

$$\left[\tilde{M}_{rr}^{*}\right] = \left[M_{rr}^{*} + \Delta M_{rr}\right] = \left[K_{rr}^{*}\right]\left[z_{rk}\right]\left[\lambda_k\right]\left[z_{rk}\right]^{-1},$$ ... (4.78)

$$\left[\Delta M_{rr}\right] = \left[K_{rr}^{*}\right]\left[z_{rk}\right]\left[1/\lambda_k\right]\left[z_{rk}\right]^{-1} - \left[M_{rr}^{*}\right].$$ ... (4.79)

The global stiffness and mass matrices for all the degrees of freedom of the system condensed to the primary levels can be obtained by the usual substructuring method by the summation of stiffness and mass matrices of individual structures.

As already pointed out, eqn. (4.75) permits different solutions, i.e., constructions of condensed subsystems equivalent to the original ones for the first n eigenvalues and corresponding components of the eigenvectors associated with them:

a) instead of matrix M_{rr}^* obtained by static condensation according to eqn. (4.12), matrix M_{rr}^{**} obtained using eqn. (4.18) with allowance for only the inertial forces is used in eqns. (4.78) and (4.79):

$$\left[M_{rr}^{**} \right] = M_{rr} - M_{rs} M_{ss}^{-1} M_{sr} ; \qquad \qquad \text{... (4.80)}$$

b) instead of matrices K_{rr}^* and M_{rr}^* in eqns. (4.76) to (4.79) matrices K_{rr}^{**} and M_{rr}^{**} derived from eqns (4.18) and (4.80) are used:

$$\left[K_{rr}^{**} \right] = K_{rr} - K_{rs} M_{ss}^{-1} M_{sr} - M_{rs} M_{ss}^{-1} M_{ss} + M_{rs} M_{ss}^{-1} K_{ss} M_{ss}^{-1} M_{sr} ; \text{ and ... (4.81)}$$

c) equation (4.75) is solved relative to the condensed stiffness matrix

$$\left[\tilde{K}_{rr} \right] = \left[\tilde{M}_{rr} \right] \left[Z_{rs} \right] \left[\lambda_s \right] \left[Z_{rs} \right]^{-1}, \ \left[\Delta K_{rr} \right] = \left[\tilde{K}_{rr} - K_{rr}^* \right]. \qquad \text{... (4.82)}$$

When condition (c) is not satisfied, the arrangement of condensation points is unsatisfactory as they fall at the zero points of some natural form of oscillation of a reduced spectrum or close to them and do not permit obtaining eigenvalues corresponding to that form. The eigenvalue calculated in such a case usually corresponds to the next higher natural frequency of the spectrum for which the zero points of the natural form no longer coincide with the condensation nodes. An illustration of this is the example of analysis whose results are shown in Table 4.9.

Example 4.8. A square thin plate 8 × 8 m in size and thickness $h = 0.05$ m with uniformly distributed mass $\overline{m} = 1$ (kg/cm²) is rigidly fixed all along the contour.

The solution to eqn. (4.6) or (4.7) for the initial FE grid of 8 × 8 was adopted as the 'accurate' solution (the finite element grid, the arrangement of estimated points and substructures are shown in Fig. 4.11).

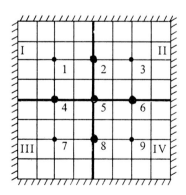

Fig. 4.11

Table 4.16 gives the results using the algorithm given above. Its characteristic feature, as pointed out, is the arbitrary selection or construction of the condensed stiffness matrix $\tilde{K}_{rr}$ for each subsystem. For a numerical experiment to confirm this feature of the algorithm, the next variant for constructing the condensed stiffness matrix $\tilde{K}_{rr}$ of the system was used. The entire system (plate in this case) was divided into individual regions (substructures). Later, static condensation of the stiffness matrix was carried out region-wise for the entire structure to the main degrees of freedom prevailing in the region covered by the given substructure. All the stiffness matrices obtained thereby were combined by the substructuring method into one global matrix K_{rr}. Since each stiffness submatrix component is statically equivalent to the stiffness matrix of the entire system, the global matrix $\tilde{K}_{rr}$ in this case is applicable to a nominal system with t times greater stiffness (t is the number of substructures).

Table 4.16

Eigenvalue	Adopted accurate solution by finite element method in 8 × 8 mesh	Static conden- sation	Δ, %	Dynamic conden- sation	Δ, %
λ_1	1.429	1.53	7.0	1.398	-2.1
λ_2	0.3483	0.369	6.1	0.348	0.0
λ_3	0.3483	0.369	6.1	0.348	0.0
λ_4	0.1657	0.175	5.4	0.163	-1.8
λ_5	0.1068	0.105	-1.7	0.107	0.0
λ_6	0.1056	0.103	-2.5	0.106	0.4
λ_7	0.0727	0.76	4.6	0.0704	-3.1
λ_8	0.0727	0.76	4.6	0.0704	-3.1
λ_9	0.0428	0.039	-9.5	0.417	-2.3

A comparison of results obtained by this dynamic condensation procedure with the results using the static condensation method showed the significantly better accuracy of the former.

Example 4.9. Rectangular plate 8 × 12 m in size and thickness $h = 0.05$ m fixed firmly along the contour and carrying uniformly distributed mass $m = 1$ (kg/cm²).

The analysis was made by the following condensation methods:

1) dynamic condensation (example 4.8);

2) statically condensed stiffness matrices K_{rr}^* of the subsystems using eqn. (4.12) and mass matrices of subsystems condensed using eqn. (4.78); and

3) usual static condensation.

The resultant analyses are shown in Table 4.17. Column 5 gives the results on dividing the system into four subsystems and column 7 the results on dividing into six subsystems.

$t = 4$ (4 substructures) $t = 6$ (6 substructures)

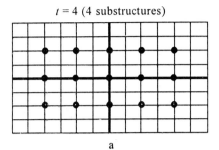
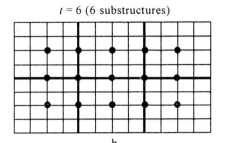

a

b

Fig. 4.12

Table 4.17

Eigen-value	Finite element method in 8×12 mesh, $n = 231$	Static conden-sation, $n = 15$	Δ, %	Dynamic conden-sation, $t = 4$	Δ, %	Dynamic conden-sation, $t = 6$	Δ, %
1	2	3	4	5	6	7	8
λ_1	2.512	2.66	6	2.372	−5	2.370	−5
λ_2	1.030	1.49	30	1.060	−2	1.051	−2
λ_3	0.423	0.479	13	0.426	0.7	0.429	1
λ_4	0.412	0.432	4.5	0.395	4	0.398	4
λ_5	0.306	0.390	27.4	0.295	−3	0.291	−4
λ_6	0.190	0.212	11.6	0.192	1	0.188	−1
λ_7	0.182	0.197	8.2	0.178	−2	0.174	−4
λ_8	0.116	0.181	56.0	0.109	−6	0.112	−3
λ_9	0.110	0.154	40.0	0.104	−6	0.101	−8
λ_{10}	0.101	0.132	20.8	0.101	0	0.093	−8
λ_{11}	0.087	0.119	34.2	0.081	−7	0.085	−2.5
λ_{12}	0.080	0.103	26.3	0.074	−7	0.081	−1
λ_{13}	0.062	0.089	43.5	0.058	−7	0.058	−7
λ_{14}	0.058	0.081	40.0	0.052	−22	0.052	−10
λ_{15}	0.045	0.063	40.0	0.036	−20	0.036	−20

Table 4.18 shows the results of analysis of the same plate by the first condensation method on dividing the system into 4 and 6 subsystems of the same FE mesh of 8×12 m. A comparison of the results obtained for different numbers of subsystems but with the same number of selected nodes (primary degrees of freedom) reveals their close proximity. This indirectly confirms the above statement that the accuracy of the frequency-dynamic condensation method depends on the ratio n/N and not on the number of subsystems.

Table 4.18

Eigenvalue	Division into 4 subsystems ($t = 4$)	Δ, %	Division into 6 subsystems ($t = 6$)	Δ, %
λ_1	2.52	1.7	2.545	6.3
λ_2	1.12	8.0	1.13	9.0
λ_3	0.46	8.7	0.456	7.9
λ_4	0.398	−8.9	0.384	−6.9
λ_5	0.307	0.3	0.301	−1.6
λ_6	0.189	−0.5	0.188	−1.0
λ_7	0.180	−1.0	0.181	−0.5
λ_8	0.111	−4.3	0.111	−4.3
λ_9	0.109	−0.9	0.108	−1.8
λ_{10}	0.100	−1.0	0.102	1.0
λ_{11}	0.083	−4.5	0.085	−2.2
λ_{12}	0.077	−3.7	0.076	−5.0
λ_{13}	0.053	−7.0	0.058	−7.0
λ_{14}	0.052	−12.0	0.052	−12.0
λ_{15}	0.036	−20.0	0.036	−20.0

A comparison of the errors of the first two variants with the 'accurate' uncondensed solution suggests their near total agreement. This confirms the above hypothesis about the arbitrary selection (or construction) of the stiffness matrix of the subsystem $\tilde{K}_{rr}$ when solving eqn. (4.75).

The static condensation method in this case is far less accurate than either of the dynamic condensation procedures.

4.12 Energy Form of Consecutive Frequency-dynamic Condensation Method

4.12.1 Condensation Using Minimum Natural Frequency of Subsystems

1) In addition to the prior assumption regarding equal eigenvalues of the characteristic matrix of the condensed system and the corresponding number of old eigenvalues of the characteristic matrix of the original system, let us introduce the following assumptions:

a) equal components of the corresponding eigenvectors of these characteristic matrices and

b) equal kinetic energy of original and condensed systems.

Based on these assumptions, the following equation can be expressed:

$$\sum_r m_r \omega^2 z_r^2 + \sum_s m_s \omega^2 z_s^2 = \sum_r \left(m_r + m_r^{(s)} \right) \omega^2 z_r^2, \qquad \dots (4.83)$$

in which $m_r^{(s)}$ represent the condensation additives which result from reducing all masses m_s to masses m_r.

After conversion, this equation assumes the form:

$$\sum_s m_s z_s^2 = \sum_r m_r^{(s)} z_r^2 , \qquad\qquad ... (4.84)$$

or in a matrix form

$$\{z_s\}^T \lceil m_s \rfloor \{z_s\} = \{z_r\}^T \left\lceil m_r^{(s)} \right\rceil \{z_r\}, \qquad\qquad ... (4.85)$$

in which $\{z_r\}$ and $\{z_s\}$ are the subvectors of the total vector of nodal displacements $\{z\}$ pertaining to the primary and secondary degrees of freedom and $\lceil m_s \rfloor$ and $\left\lceil m_r^{(s)} \right\rceil$ are the diagonal matrices of orders s and r respectively.

Having examined the form of oscillation corresponding to the minimum fundamental natural frequency ω_{min}, let us express eqn. (4.38) in the following form:

$$\{z_s\} = \left[D_{rr}^{(s)} (\lambda_{max}) \right]\{z_r\}, \qquad\qquad ... (4.86)$$

in which

$$\left[D_{rr}^{(s)} (\lambda_{max}) \right] = -[\delta_{ss} M_s - \lambda E]^{-1} [\delta_{sr}][M_r].$$

By substituting eqn. (4.86) in (4.85), we obtain the dependence

$$\{z_r\}^T \left[D_{rr}^{(s)} (\lambda_{max}) \right]^T \lceil m_s \rfloor \left[D_{rr}^{(s)} (\lambda_{max}) \right] \{z_r\} = \{z_r\}^T \left\lceil m_r^{(s)} \right\rceil \{z_r\},$$

from which we find

$$\left\lceil m_r^{(s)} \right\rceil = \left[D_{rr}^{(s)} (\lambda_{max}) \right]^T \lceil m_s \rfloor \left[D_{rr}^{(s)} (\lambda_{max}) \right]. \qquad\qquad ... (4.87)$$

A quadratic matrix of order $r \times r$ is obtained on the right side of eqn. (4.87). When it is required to reduce it to the diagonal form corresponding to the matrix structure $\left\lceil m_r^{(s)} \right\rceil$, all the rows in it are added and the result is expressed in the form of a diagonal matrix.

By similarly finding the matrices of condensed masses for each of the t groups of masses supplied at the auxiliary nodes, we obtain the final equation for the condensed system:

$$\left| \delta_{rr} \left(M_r + M_r^{(s),t} \right) - \lambda E_r \right| = 0, \qquad\qquad ... (4.88)$$

in which

$$\left[M_r^{(s)} \right] = \sum_t \left| m_r^{(s)t} \right|.$$

Here, summation is carried out for all the t groups into which the auxiliary degrees of freedom have been divided.

2) In the partial case when the original system has been condensed to a system with a single degree of freedom, all the above equations are greatly simplified.

Let us examine initially a two-mass system (Fig. 4.13) described by the following system of equations:

$$z_r \left(m_r \delta_{rr} - \lambda \right) + z_s \delta_{rs} m_s = 0,$$

$$z_r m_r \delta_{sr} + z_s \left(m_s \delta_{ss} - \lambda \right) = 0, \qquad \ldots (4.89)$$

in which $\lambda = \omega^{-2}$.

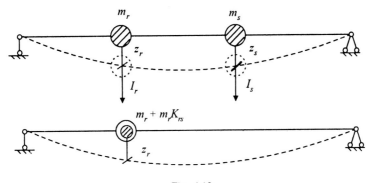

Fig. 4.13

Let us condense it to a single mass system.

From the equality conditions of the kinetic energy of the original and condensed systems described before (4.83), we find the condensation coefficient for the present case:

$$K_{rs} = \frac{z_s^2}{z_r^2} \cdot \frac{m_s}{m_r}. \qquad \ldots (4.90)$$

The dependence between oscillation amplitudes z_r and z_s emerges from the system of equations (4.89):

$$z_s = z_r \cdot \frac{\left(\lambda - m_r \delta_{rr} \right)}{m_s \delta_{rs}}, \qquad \ldots (4.91)$$

or

$$z_s = z_r \cdot \frac{m_r \delta_{sr}}{\left(\lambda - m_s \delta_{ss} \right)}. \qquad \ldots (4.92)$$

Substituting these equations in (4.90), we find

$$K_{rs} = \frac{\left(\lambda - m_r \delta_{rr} \right) m_s}{m_s \delta_{rs} m_r} \quad \text{or} \quad K_{rs} = \frac{m_r \delta_{sr} m_s}{\left(\lambda - m_s \delta_{ss} \right)}, \qquad \ldots (4.93)$$

in which λ for each of the partial systems is determined by eqn. (4.47).

The total value of the condensed mass obtained by condensing all the t masses m_s is found by usual summation of all the condensation stages:

$$M_r = m_r + \sum_{s=1}^{t} K_{rs}\, m_r = m_r \left(1 + \sum_{s=1}^{t} K_{rs}\right). \qquad \dots (4.94)$$

Example 4.10. Let us examine the same system as given in Example 4.3 of Section 4.7. Using the algorithm discussed in para 2, let us condense to mass m_3 falling at the centre of the beam span (Fig. 4.3).

We find from eqns. (4.93):

$$K_{32} = K_{34} = 0.782, \quad K_{31} = K_{34} = 0.263.$$

Substituting these results in eqn. (4.88) and solving it will give

$$\lambda = \frac{250.29}{3888} \cdot \frac{Ml^3}{EJ} = 0.064375 \frac{Ml^3}{EJ}.$$

This value of λ differs from the accurate value by 4.1%.

4.12.2 Condensation Using the Reduced Spectrum of Eigenvalues and Eigenvectors of Subsystems

Let us divide as before the displacement vector of the subsystem into two subvectors, i.e.,

$$\{z\} = \lfloor z_r,\, z_s \rfloor^T. \qquad \dots (4.95)$$

Let us express components z_s of the displacement subvector in the direction of secondary degrees of freedom as components of z_r of the displacement subvector in the direction of the primary degrees of freedom:

$$\{z_s\} = [A_s]\{z_r\}. \qquad \dots (4.96)$$

Here, $[A_s]$ is the conversion matrix whose structure will be determined below.

Equation (4.96) is applicable for any k-th form of oscillations, i.e.

$$\{z_{s,k}\} = [A_s]\{z_{r,k}\}. \qquad \dots (4.97)$$

To construct matrix $[A_s]$, we use the initial r of n eigenvectors of the subsystem and their subvectors pertaining to the primary and secondary degrees of freedom. By substituting in eqn. (4.97) the matrices of the subvectors for $z_{s,k}$ and $z_{r,k}$, we derive

$$\{z_{s,k}\} = [A_s][z_{r,k}]. \qquad \dots (4.98)$$

Since matrix $\{z_{r,k}\}$ is quadratic, matrix $[A_s]$ is determined by equation

$$[A_s] = [z_{s,k}][z_{r,k}]^{-1}. \qquad \dots (4.99)$$

Thus,

$$\{z\} = \left\{ \begin{array}{c} z_r \\ z_s \end{array} \right\} = \left\{ \begin{array}{c} z_r \\ A_s z_r \end{array} \right\} = [B]\{z_r\}, \qquad \ldots(4.100)$$

in which $[B] = \left[\begin{array}{c} E_r \\ A_s \end{array} \right]$.

Let us then express the equations for the kinetic and potential energies of the subsystem:

$$KE = \frac{1}{2} \dot{z}^T M \dot{z}, \qquad \left(\dot{z} = \frac{dz}{dt} \right)$$

$$PE = \frac{1}{2} z^T K z. \qquad \ldots(4.101)$$

Here, M and K are the mass and stiffness matrices of the subsystem respectively.

By substituting in eqn. (4.101) the dependence (4.100), conversion is made to the new co-ordinates-main degrees of freedom:

$$KE = \frac{1}{2} \dot{z}_r^T \tilde{M} \dot{z}_r, \qquad \ldots(4.102)$$

$$PE = \frac{1}{2} z_r^T \tilde{K} z_r.$$

Here, $\tilde{K}$ and $\tilde{M}$ are the stiffness and mass matrices of the subsystem condensed to the primary degrees of freedom:

$$\tilde{K} = B^T K B, \quad \tilde{M} = B^T M B. \qquad \ldots(4.103)$$

The division into subsystems, as shown in the preceding sections, can be carried out mathematically (division into blocks of coefficient matrices of the finite element method equations) and also at the physical level (division into superelements of different levels in the substructuring method). In either case, there is stagewise growth of the subsystem to the level of the system as a whole leaving in the last stage a predetermined number and position of the nodes considered (primary degrees of freedom).

The formation of global stiffness and mass matrices also proceeds in stages. On transition from the i-th to $(i+1)$-th condensation stage, the global primary nodes from the preceding subsystems are additionally included in the primary degrees of freedom of the subsystem at the $(i+1)$-th level.

When analysing in the form of forces (see Section 4.8), the condensed mass matrix of the subsystem can be constructed as follows.

Let us substitute eqn. (4.96) in the system of equations (4.37). After conversions, the first equation of this system takes the form

$$\left[\delta_{rr} (M_r + \Delta M_r) - \lambda E_r \right] \{z_r\} = 0, \qquad \ldots(4.104)$$

in which the following notation was used:

$$\delta_{rr}\,\Delta M_r = \Delta\big(\delta_{rr}\,M_r\big) = \delta_{rs}\,M_s\,A_s.\qquad\ldots(4.105)$$

Here, ΔM_r is the matrix of additives condensed to matrix M_r.
It follows from equation (4.105) that

$$\Delta M_r = \delta_{rr}^{-1}\,\delta_{rs}\,M_s\,A_s.\qquad\ldots(4.106)$$

Two rectangular plates of constant thickness with an initial mesh of quadratic finite elements 4 × 8 (Table 4.19) and 4 × 12 (Table 4.20) and firmly fixed along the contour were analysed using the above condensation algorithm in the form displacement method. The plate mass was assumed as uniformly distributed: $\overline{m} = 1$.

Table 4.19

Finite element grid and arrangement of condensation nodes	No. of condensation nodes	100 × n/N, %	No. of eigenvalue, λ	Accurate value, λ	Static condensation, λ (Δ%)	Stepwise consecutive frequency dynamic condensation, λ (Δ%)
1	2	3	4	5	6	7
	3	5	1	0.1919	0.177 (−7.7)	0.1910 (−0.5)
			2	0.1211	0.088 (−27.3)	0.1194 (−1.4)
			3	0.0527	0.032 (−39.3)	0.448 (−15)
	3	5	1	0.1919	0.177 (−7.7)	0.1910 (−0.5)
			2	0.1211	0.088 (−27.3)	0.1194 (−1.4)
			3	0.0527	0.032 (−39.3)	0.0448 (−15)
	3	5	1	0.1919	0.177 (−7.7)	0.1910 (−0.5)
			2	0.1211	0.088 (−27.3)	0.1194 (−1.4)
			3	0.0527	0.032 (−39.3)	0.0448 (−15)
	3	5	1	0.1919	0.177 (−7.7)	0.1910 (−0.5)
			2	0.1211	0.088 (−27.3)	0.1194 (−1.4)
			3	0.0527	0.032 (−39.3)	0.0448 (−15)
	5	8	1	0.1919	0.177 (−7.7)	0.1889 (1.5)
			2	0.1211	0.088 (−27.3)	0.1219 (0.7)
			3	0.0527	0.032 (−39.3)	0.0552 (4.7)

Contd.

Table 4.19 Contd.

1	2	3	4	5	6	7
			4	0.0310	0.026 (−16.1)	0.0284 (−8.4)
			5	0.0276	0.019 (−31.1)	0.0261 (−5.4)
	5	8	1	0.1919	0.177 (−7.7)	0.1923 (0.2)
			2	0.1211	0.088 (−27.3)	0.1220 (0.7)
			3	0.0527	0.032 (−39.3)	0.0555 (5.3)
			4	0.0310	0.026 (−16.1)	0.0284 (−8.4)
			5	0.0276	0.019 (−31.1)	0.0262 (−5.2)
	5	8	1	0.1919	0.177 (−7.7)	0.1929 (0.5)
			2	0.1211	0.088 (−27.3)	0.1219 (0.7)
			3	0.0527	0.032 (−39.3)	0.0555 (5.3)
			4	0.0310	0.026 (−16.1)	0.0284 (−8.4)
			5	0.0276	0.019 (−31.1)	0.0261 (−5.4)
	5	8	1	0.1919	0.177 (−7.7)	0.1926 (0.36)
			2	0.1211	0.088 (−27.3)	0.1218 (0.6)
			3	0.0527	0.032 (−39.3)	0.0555 (5.3)
			4	0.0310	0.026 (−16.1)	0.0284 (−8.4)
			5	0.0276	0.019 (−31.1)	0.0261 (−5.4)
	5	8	1	0.1919	0.177 (−7.7)	0.1926 (0.36)
			2	0.1211	0.088 (−27.3)	0.1917 (0.5)
			3	0.0527	0.032 (−39.3)	0.0554 (5.1)
			4	0.0310	0.026 (−16.1)	0.0284 (−8.4)
			5	0.0276	0.019 (−31.1)	0.0261 (−5.4)
			1	0.1919	0.177 (−7.7)	0.1926 (0.3)
			2	0.1211	0.088 (−27.3)	0.1218 (0.6)

Contd.

Table 4.19 Contd.

1	2	3	4	5	6	7
	5	8	3	0.0527	0.032 (−39.3)	0.0554 (5.1)
			4	0.0310	0.026 (−16.1)	0.0284 (−8.4)
			5	0.0276	0.019 (−31.1)	0.0262 (−5.4)
			1	0.1919	0.177 (−7.7)	0.1925 (0.3)
			2	0.1211	0.088 (−27.3)	0.1212 (0.1)
	5	8	3	0.0527	0.032 (−39.3)	0.0553 (4.9)
			4	0.0310	0.026 (−16.1)	0.0284 (−8.4)
			5	0.0276	0.019 (−31.1)	0.0261 (−5.4)
			1	0.1919	0.177 (−7.7)	0.1926 (0.36)
			2	0.1211	0.088 (−27.3)	0.1218 (0.6)
	6	10	3	0.0527	0.032 (−39.3)	0.0555 (5.3)
			4	0.0310	0.026 (−16.1)	0.0286 (−7.7)
			5	0.0276	0.019 (−31.1)	0.0271 (−1.8)
			6	0.0241		0.0261 (8.3)
			1	0.1919	0.177 (−7.7)	0.1930 (0.57)
			2	0.1211	0.088 (−27.3)	0.1218 (0.6)
			3	0.0527	0.032 (−39.3)	0.0556 (5.4)
			4	0.0310	0.026 (−16.1)	0.0296 (−4.5)
	10	16	5	0.0276	0.019 (−31.1)	0.0274 (−0.7)
			6	0.0241		0.0261 (8.3)
			7	0.0189		0.0192 (1.5)
			8	0.0160		0.0180 (12.5)
			9	0.0135		0.0160 (18.5)
			10	0.0090		0.0124 (37.7)

Note. The percentage deviation from the solution of the finite element method adopted as accurate is shown within parentheses.

Table 4.20

Finite element grid and arrangement of condensation nodes	No. of condensation nodes	$100 \times n/N$, %	No. of eigenvalue, λ	Accurate value, λ	Static condensation, λ	Stepwise consecutive frequency dynamic condensation, λ
1	2	3	4	5	6	7
	5	5	1	0.2105	0.2020 (−4.0)	0.2108 (0.14)
			2	0.1760	0.1480 (−18.0)	0.1735 (−1.4)
			3	0.1295	0.0853 (−34.1)	0.1295 (0)
			4	0.0863	0.0401 (−53.5)	0.0668 (−22.3)
			5	0.0548	0.0274 (−50.0)	0.0360 (−34.3)
	8	8	1	0.2105	0.2020 (−4.0)	0.2108 (0.14)
			2	0.1760	0.1480 (−18.0)	0.1780 (1.1)
			3	0.1295	0.0853 (−34.1)	0.1295 (0)
			4	0.0863	0.0401 (−53.5)	0.0834 (−3.3)
			5	0.0548·	0.0274 (−50.0)	0.0360 (−34.3)
			6	0.0344	0.0232 (−32.5)	0.0267 (−22.3)
			7	0.0283	0.0172 (−39.2)	0.0210 (−25.8)
			8	0.0268	0.0154 (−42.5)	0.0154 (−42.5)
	10	10	1	0.2105	0.2020 (−4.0)	0.2108 (0.14)
			2	0.1760	0.1480 (−18.0)	0.1780 (1.1)
			3	0.1295	0.0853 (−34.1)	0.1295 (0)
			4	0.0863	0.0401 (−53.5)	0.0839 (−3.8)
			5	0.0548	0.0274 (−50.5)	0.0525 (−6.1)
			6	0.0344	0.0232 (−32.5)	0.0269 (−21.0)
			7	0.0283	0.0172 (−39.2)	0.0210 (−25.8)
			8	0.0268	0.0154 (−42.5)	0.0154 (−42.5)
			9	0.0244		0.0139 (−43.0)
			10	0.0219		0.0064 (−300)

Note. The percentage deviation from the solution of the finite element method adopted as accurate is shown within parentheses.

Tables 4.19 and 4.20 give the results of analysis of the first and second plates respectively for different arrangements of condensation nodes and compare the results using the proposed algorithm and applying the static condensation method.

Analysis of results given in Table 4.19 leads to the following conclusions:

(1) the arrangement of condensation nodes does not significantly influence the accuracy of results and

(2) analysis by the proposed algorithm ensures a significant improvement in accuracy compared to the static condensation method.

Analysis of results given in Table 4.20 makes it possible to conclude that, compared to the static condensation method, the proposed algorithm gives results very close to the accurate ones for 50 to 60% of the eigenvalues in the reduced spectrum, while the static condensation method gives acceptable results only for the maximum eigenvalue.

4.13 Generalisation of Consecutive Frequency-dynamic Condensation Method in Stability Problems

It is known (Maslenninov, 1987; Chapter 2) that the energy method can form a basis for analysing structural stability. According to the concept of this method, we obtain the following equation in a matrix form for a neutral state of equilibrium of the structure:

$$Kz = Yz = 0 \qquad \qquad ...(4.107)$$

in which K is the stiffness matrix of the system, z the displacements matrix and Y the matrix of the loading potential.

Since $z \neq 0$ when there is loss of stability and the entire load is assumed as proportional to one single parameter P, eqn. (4.107) leads to the following form:

$$\left| K - PY_0 \right| = 0. \qquad \qquad ...(4.108)$$

On introducing the notation $\lambda = 1/P$ and $C = K^{-1}Y_0$, eqn. (4.111) takes the standard form for problems involving eigenvalues and eigenvectors:

$$\left| C - \lambda E \right| = 0. \qquad \qquad ...(4.109)$$

Matrix Y is determined by summation of the work of external forces of individual FE (when analysing by the finite element method):

$$Y = \sum_{g=1}^{n} \left(a_g^T Y_g a_g \right) \qquad \qquad ...(4.110)$$

in which Y_g is the matrix of the loading potential of individual elements depending on the form of the stability loss of FE and type of loading; and a_g the deformation matrix of individual FE due to unit nodal displacements.

For example, for an FE in the form of a rectangular plate, an incomplete cubic polynomial can be adopted as a function of displacements when there is loss of stability as in the case of deriving the stiffness matrix of an FE:

$$z_g(x,y) = \alpha_1 + \alpha_2 x + \alpha_3 y + \alpha_4 x^2 + \alpha_5 y^2 + \alpha_6 xy + \alpha_7 x^2 y +$$

$$\alpha_8 xy^2 + \alpha_9 x^3 + \alpha_{10} y^3 + \alpha_{11} x^3 y + \alpha_{12} xy^3 = \left[\Phi^{(g)}(x,y) \right] \{\alpha^g\}. \qquad \dots (4.111)$$

Then, the work of external forces for different FE represented in a matrix form, will assume the form

$$Y_g = \int_s \left[\frac{dz}{ds} \right]^T P \left[\frac{dz}{ds} \right] ds, \qquad \dots (4.112)$$

in which

$$\left[\frac{dz}{ds} \right] = \left\{ \begin{array}{c} \frac{dz}{dx} \\ \frac{dz}{dy} \end{array} \right\} = \begin{bmatrix} 0 & 1 & 0 & 2x & 0 & x & 2xy & y^2 & 3x^2 & 0 & 3x^2 y & y^3 \\ 0 & 0 & 1 & 0 & 2y & y & x^2 & 2xy & 0 & 3y^2 & x^3 & 3xy \end{bmatrix} \times$$

$$\{\alpha^g\} = [G]\{\alpha^g\}. \qquad \dots (4.113)$$

Here, as for a stiffness matrix

$$\{\alpha^g\} = [V]^{-1}\left[\alpha^g\right] \qquad \dots (4.114)$$

in which $[V]$ is the matrix obtained on the basis of eqn. (4.111) after substituting for x and y the nodal co-ordinates of finite elements (Maslennikov, 1987; Chapter 2); and $\{z^g\}$ the vector of nodal displacements of the FE.

On substituting eqns. (4.107) and (4.111) in (4.112), we obtain the matrix of the loading potential for a given finite element:

$$Y_g = \int_s (V^{-1})^T G^T PG(V^{-1}) ds. \qquad \dots (4.115)$$

On integration for a given specific matrix P, we obtain the value of matrix elements Y_g.

To determine the spectrum of critical forces and the corresponding forms of stability loss, the method of consecutive dynamic condensation proposed above can be applied. After separating the degrees of freedom into main (r) and slave (s) eqn. (4.108) can be represented in a block form as:

$$\left(\begin{bmatrix} K_{rr} & K_{rs} \\ K_{sr} & K_{ss} \end{bmatrix} - \lambda \begin{bmatrix} Y_{rr} & Y_{rs} \\ Y_{sr} & Y_{ss} \end{bmatrix} \right) \left\{ \begin{array}{c} z_r \\ z_s \end{array} \right\} = 0. \qquad \dots (4.116)$$

By subdividing the auxiliary unknowns into t groups and eliminating them blockwise, we derive the characteristic equation for the reduced system:

$$\left(K_{rr} - \lambda Y_{rr}\right) + \left(\sum_{s=1(s \neq r)}^{m} K_{rr}^{(s)} - \lambda \sum_{s=1}^{m} Y_{rr}^{(s)}\right) = 0. \qquad \dots (4.117)$$

To evaluate the accuracy of analysis by the proposed algorithm using the approximating function (4.114), an analysis was made of the stability of a plate hinged along the contour (Fig. 4.14) against the impact of compressive loading uniformly distributed along the contour.

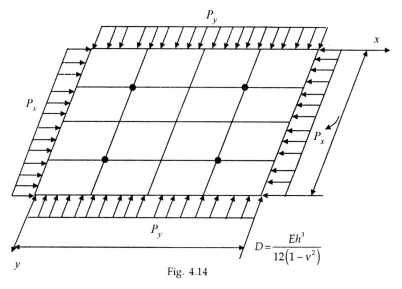

Fig. 4.14

The finite-element grid of the plate was divided into 16 FE. The primary nodes r are shown as dots in the Figure. The plate parameters have been given in Section 4.4 (Fig. 4.1). Table 4.21 shows the results of comparing the spectra of critical forces obtained for the initial grid of FE with the results obtained for the condensed system and with the accurate solution (Bezukhov et al., 1987; Chapter 2).

Table 4.21

No.	Accurate value of P_{cr}	Finite element method, 4 × 4	Δ, %	Dynamic condensation ($n = 4$)	Δ, %	Δ, % of finite element method
P_1	1.6449	1.609	−2.18	1.6025	−2.58	−0.4
P_2	4.1123	3.9866	−3.06	3.8733	−5.8	−2.84
P_3	4.1123	3.9866	−3.06	3.8733	−5.8	−2.84
P_4	6.5795	6.1292	−6.8	6.3292	3.82	3.26

Common multiplier D/l^2.

The time required for analysis by computer using the proposed algorithm was reduced to one-fifth that for the original problem (4.116) (Blokhina, 1989; Chapter 2).

References

*Argiris, J.H. 1961. *Modern Methods of Analysing Complex Statically Indeterminate Systems.* Sudpromgiz, Leningrad, 690 pp.

*Belyi, M.V. and T.N. Gorbunova. 1993. *Multilevel Iteration Algorithms for Frequency and Form Analysis of Natural Oscillations of Structures.* Moscow State Engineering University, Moscow, pp. 1–13.

*Blokhina, I.V. 1987. Efficiency comparison of various dynamic condensation methods when determining the frequencies of natural oscillations of complex rod systems. Volgograd Engineering-Structural Institute, Volgograd. Deposited in VINITI, 3 Feb. 1987, No. 1134, 25 pp.

Bouhaddi, N., Fillod, R. Model reduction by a simplified variant of dynamic condensation. Joural of Sound and Vibration V191 n2 Mar 28 1996. P233-250.

Bouhaddi, N., Fillod, R. Substructuring by two level dynamic condensation method. Computers and Structures v60n3 Aug 3 1996. p403-409.

Craig, R.R. and M.C.C. Bampton 1968. *American Institute of Aeronautics Journal* 6, 1313-1319. Coupling of substrctures for dynamic analysis.

*Craig, R.R. Jr. and Chaing-Jone Chang. 1976. Free interface methods of substructure coupling for dynamic analysis, *AIAA J.*, 14(11): 1633–1635. *Raketnaya Tekhnika i Kosmonavtika*, 14(11): 154–156.

Dieker, S. 1991. Ein verallgemeinertes Verfahren der Substrukturtechnik fuer den Aufbau dynamischer Superelemente. *Zeitschrift fuer Flugwissenschaften und Weltraumforschung*, 15 (1991), pp. 379–385. Springer-Verlag.

Egeland, O., Azaldsen, P.O. SESAM-69-a general purpose finite element program.-Int. J. Computers and Structures, 1974, 4.

*Grinenko, N.I. and V.V. Mokeev. 1985. Problems of studying structural oscillations by the finite element method. *Prikladnaya Mekhanika*, 21(3): 12–15.

*Grinenko, N.I. and V.V. Mokeev. 1988. Improving the efficiency of the finite element method in problems of designing dynamic systems. In: *Analysis and Control of the Reliability of Large Mechanical Systems.* Sverdlovsk-Tashkent, pp. 20–25.

*Guyan, R.J. 1965. Reduction of stiffness and mass matrices. *AIAA J.*, 3(2): 380.

*Ignateev, V.I. and V.D. Chuban. 1984. Form and frequency analysis of free oscillations in structures by multilevel dynamic condensation method. *Uchenye Zapiski TsAGI*, 15(4): 81–92.

*Ignatiev, V.A. 1979. *Analysis of Periodic Statically Indeterminate Rod Systems.* Saratov University, Saratov, 295 pp.

*Ignatiev, V.A. 1992. *Reduction Methods of Analysis in Statics and Dynamics of Plate Systems.* Saratov University, Saratov, 142 pp.

Kammer D.C., C.C. Flanigan. Development of test-analysis models for large space structures using substructure representations. J. Spacecraft, Vol. 28, No. 2, March-April 1991. pp.244-250.

Ki-Ook Kim. 1974. Hybrid dynamic condensation for eigenproblems. Computers and Structures. Vol. 56 pp. 105-112, 1995.

Ki-Ook Kim and J. Anderson. 1984. Generalised dynamic reduction in finite element dynamic optimisation. *AIAA J.*, 22(11): 1616–1617.

Kim, K.O. 1985. Dynamic condensation for structural re-design, *AIAA J.* 23, 1830-1831.

Leung. A.Y.T. 1978. An accurate method of dynamic condensation in structural analysis. *Int. J. Numer. Meth. Engg.* 12, 1705-1715.

Leung, A.Y.T. 1989. Dynamic stiffness and non-conservative modal analysis. *Intern. J. Anal. Exper. Mech.* pp. 77–82.

Leung A.Y.T. 1993. *Dynamic Stiffness and Substructures.* Berlin: Springer-Verlag.

Levy, C. 1991. An iterative technique based on the Dunkerley method for determining the natural frequencies of vibrating systems. *J. Sound and Vibration*, 150(1): 111–118.

*Nemchinov, Yu.I. and A.A. Kozyr'. 1985. Application of the condensation of dynamic variables in the spatial finite element method. *Stroitel'naya Mekhanika i Raschet Sooruzhenii*, 1: 51–54.

Paz, M. 1984. Dynamic condensation. *AIAA J.* 22, 724-727.

Rosman, R. 1990. Naeherungsweise Loesung von Eigenwertaufgaben der Baumechanik durch Aufspaltung in Teilaufgaben. *Baumechanik*, 67(11): 375–382.

Suarez, L.E. and M.P. Singh. 1992. Dynamic condensation method for structural eigenvalue analysis. *AIAA J.* 30, 1045-1054.

Singh, M.P., Suarez L.E. 1992. Dynamic condensation with synthesis of substructure eigenproperties. Journal of Sound and Vibration. 159(1), p.139-155.

Tokuda, N. and Y. Sakata. 1989. Application of static condensation method to vibration analysis of thin cylindrical shells. *J. Pressure Vessel Tech.*, 111: 275–284.

Zemljanuchin, S.J. (Zemlyanukhin, S.Ya.). 1960. Vereinfachte Berechnung der Eigenschwingungsfrequenz von Spindeln an Werkzeugmaschinen. *Industrie-Anzeiger*, Essen, 82, Jahrgang Nr. 10.

*Zemlyanukhin, S.Ya. 1962. Flexural Vibrations of Shafts of Variable Section. Cand. Diss. Technical Sciences: 01.02.03. Saratov Polytechnical Institute, Saratov, 98 pp.

*Zemlyanukhin, S.Ya. 1963. Improving the method of determining oscillation frequency and solving stability problems. *Stroitel'naya Mekhanika i Raschet Sooruzhenii*, 4: 42–46.

*Zenkevich, O. 1975. *Finite Element Method in Technology.* Mir, Moscow, 544 pp.

Zhang, Y. and R.S. Harichandran. 1989. Eigenproperties of large-scale structures by finite element partitioning and homotopy continuation. *Intern. J. Numerical Methods Engineering*, 28: 2133-2122.

*Zhuravleva, A.M. and V.G. Gadyaka. 1989. Dynamic condensation of finite element equations for oscillations in plate structures by form synthesis method. In: *Dynamics and Strength of Machines*, Khar'kov State University, Khar'kov, 49: 70–77.

Zienkiewicz, O.C. 1967. "The Finite Element Method in Structural and Continuum Mechanics", McGraw-Hill, London.

---

*Asterisked References are in Russian—General Editor.